Environmental Analysis

Environmental Analysis

Chandra Mani

Oxford Book Company
Jaipur ▪ New Delhi

Published By

Oxford Book Company

Regd. Off. "Bony Residency", Gate No. 2
Opp. Tilak Public School, Vishweswariya Nagar,
Gopalpura Road, Jaipur - 302018, India
Ph: 0141 - 2761280, Telefax: 2761381
email - oxfordbookjpr@gmail.com
website - oxfordbookcompany.com

Delhi Office: 102, 1st Floor, Satyam House,
4327/3, Ansari Road, Darya Ganj,
New Delhi - 110002, India
Ph: 011-45652440

Published in India, 2025

ISBN : 9789380179667

Price : ₹ 2550/-

Typeset by Shivangi Computers, Jaipur

Printed and bound in India by Thomson Press Ltd.,
New Delhi

Preface

Ecological research is not only confined to mere qualitative approach but is more confined to quantitative approach. In order to have a comprehensive evaluation of an ecosystem, it becomes essential to monitor both qualitatively and quantitatively spatial and temporal variations in structural and functional attributes of environment which may be categorized as physical, chemical and biological. Environmental pollution control is a major concern and challenge for twenty first century people. A comprehensive manual on environmental analytical chemistry to cater to the needs of environmental science students and researchers is a long felt need.

In spite of the intense efforts undertaken in recent years, our knowledge of the modes of distribution of elements within the biosphere and lithosphere has been still in a comparatively primitive state. While carrying out environmental analysis a scientist must be able to answer various questions. Environmental Monitoring and Analysis, an attempt has been made to describe various methods of monitoring and analysis of air, noise and water. The present book is designed to cover all important issues relating to the subject.

Chandra Mani

Contents

Chapter 1

The Model of the Earth

THE INTERIOR OF THE EARTH AS REVEALED BY EARTHQUAKES

The deep interior of the earth is inaccessible, and no rays of light penetrate to let us see what is below the surface. But rays of another kind penetrate and carry with them their messages from the interior. The Earth has been found to have elastic properties that allow movement set up at the source (focus) of an earthquake to radiate into the interior and to spread over the surface. In a strong earthquake the whole of the Earth is set vibrating.

At some distance from the focus, depending on the strength of the shock, the movement is no longer perceptible, but sensitive seismographs can record the waves that emerge at the surface. The records provide data from which knowledge of the Earth's interior may be gained. From seismic studies we have learned that the Earth consists of a core surrounded by a mantle on which there is a crust; inside the core there is a small inner core.

Towards the end of the nineteenth century it was realized that earthquake movements extend to great distances from the focus. Systematic recording of earthquakes began, and the very important results gained at an early stage greatly stimulated interest in this new science. The elastic waves that radiate into the Earth are of two kinds, having different speeds of travel: P waves (*undae primae*), in which the particle motion is longitudinal, and S waves with transverse particle motion.

The speed of S waves is roughly 60 per cent of that of P waves. The arrivals of the waves are marked by groups of oscillations in the seismograms, and at moderate distances from the focus they usually stand out clearly. The seismogram obtained at a distance of 18-61 from the focus. P and S appear on it. The large oscillations succeeding S are due to surface waves.

We measure the arrival times of the waves today usually with an accuracy of not less than 1 second, and deduce the travel times of the waves from focus to recording station, if the location of the focus and the time of occurrence of the earthquake can be found.

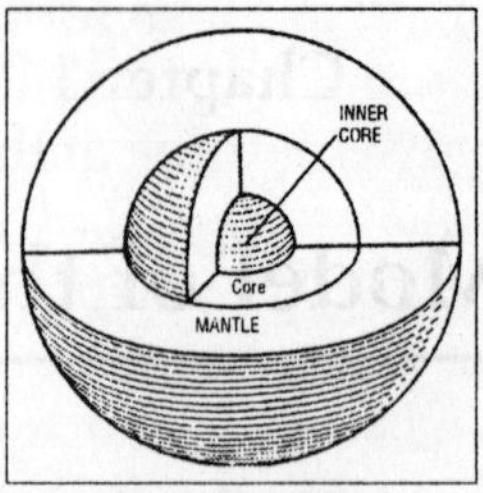

Fig. Mantle, Core, and Inner Core of the Earth

When a fair number of travel times to points at different distances from foci at approximately the same depth (the normal depth is about 10-20 km) are available, time distance tables can be set up, giving the travel times of each type of wave over the various distances. It is customary to present them as time-curves.

The first attempts to construct time-curves for the P and S phases revealed that the average speed, as determined along the chords, increased with focal distance, indicating the wave velocity increased with depth. As a consequence of this the rays are not straight lines but have an upward curvature.

At about 100° focal distance P and S became small and at somewhat greater distances could not be detected. Clear P phases reappeared at about 140°, but they had been delayed, so that their travel times did not fit on to the continuation of the time-curve for shorter distances. This observation was made in 1906 by R. D. Oldham, who drew the conclusion that deep in the Earth there was a decrease of velocity, causing the rays to bend downwards so as to leave part of the Earth's surface in shadow.

He attempted to calculate the depth at which the decrease of velocity occurred, but his observations were not good enough for this purpose; therefore his results were very much in error. E. Wiechert, in Göttingen, had also come to the conclusion that the Earth had a core in which the velocity was smaller than in the surrounding mantle. It was the first great achievement of B. Gutenberg to establish this beyond doubt by means of records of distant earthquakes obtained on the Wiechert seismographs at Gottingen, and to calculate the radius of the core. His result did not differ much from later ones obtained from modern and more abundant data.

Gutenberg made use of the time-curves for P and S up to about 103° established by Wiechert and Zöppritz, and of the wave velocities derived from them. At that time formulae had been developed by means of which transmission times to different distances could be calculated when the velocity as a function of the distance from the Earth's centre was known, and also formulae by means of which the wave velocities as a function of depth could be obtained when the time-curve was known. The ray emerging at the distance where the P curve broke off would graze the core.

The belated P waves observed from a little beyond 140° onwards had passed through the core, but the velocity of these waves in the core could not be derived from the formulae used for the mantle, for these formulae break down when there is a discontinuous decrease of velocity, as at the core boundary. But when a velocity distribution in the core was assumed the travel times could be calculated and Gutenberg varied his assumptions until the calculated travel times agreed with those observed. The velocity as a function of depth was then known for the whole of the Earth to a first approximation.

On theoretical grounds it was to be expected that the rays would be reflected on reaching the surface of the Earth and would be reflected and refracted at discontinuity surfaces in the interior, such as the boundary of the core, partly as rays of the same kind and partly transformed into rays of the other kind.

Thus at great distances from the epicentre (the point on the Earth's surface directly above the focus) waves would be arriving along many different parths, producing oscillations in a seismogram and marking phases more or less prominent according to the energy, carried. When the velocity distribution within the Earth is known, it is possible to calculate the transmission times along all the different paths.

When earthquake records are examined a great many of the anticipated phases can be identified but some of those originally expected to be present are not found, namely all those that would have come as S waves through the core. Since phases may be present without being very clear, many years passed before it was definitely concluded that transverse waves were not transmitted through the core. At the surface of the Earth a fluid does not transmit transverse waves, and therefore we say that the core is fluid, although in other respects it may not resemble a fluid as we know it.

The shadow zone for the P phase extends from about 105° to 143° epicentral distance. With modern highly sensitive instruments the P phase is found not to be completely absent in this range: in strong earthquakes it is usually faintly recorded. The appearance of P waves in the shadow zone may be due either to diffraction around the core boundary or to a spreading of the rays caused by a small gradual decrease of velocity just outside the core. In addition to this faint P phase there is, in the shadow zone, another later P phase that is faint at the smaller distances. On Gutenberg's original Earth model its presence could not be explained and it was vaguely ascribed to diffraction. However, as seismographs improved it was more and more clearly recorded, and an explanation was required.

In 1936 the writer pointed out that the presence of a small inner core in which the P velocity was greater than just outside it would fully account for the occurrence of the phase, for it would cause enough incident rays to bend upwards strongly enough for part of them to emerge in the shadow zone.

Gutenberg and Richter accepted the existence of this inner core and calculated its radius and the velocity distribution in it. Later H. Jeffreys proved that diffraction could not account for the phase in question flected once and twice at the surface of the Earth. PS starts as P and is reflected as S is part of a seismogram recorded at an epicentral distance of 108°.

P is here the P wave reflected at the boundary of the inner core. SKS starts as an S wave, traverses the core as a P wave, and is again an S wave after leaving the core. SKKS is also an S wave outside the core and a P wave inside, but it is reflected when, from inside, it meets the core boundary.

The wave velocity in the crust is smaller than in the mantle underneath, and therefore the waves coming through the crust are refracted and bent upwards when they meet the mantle. There will be a range of distance within which both the refracted and the direct waves emerge at a certain distance the refracted wave overtakes the direct wave because it travels

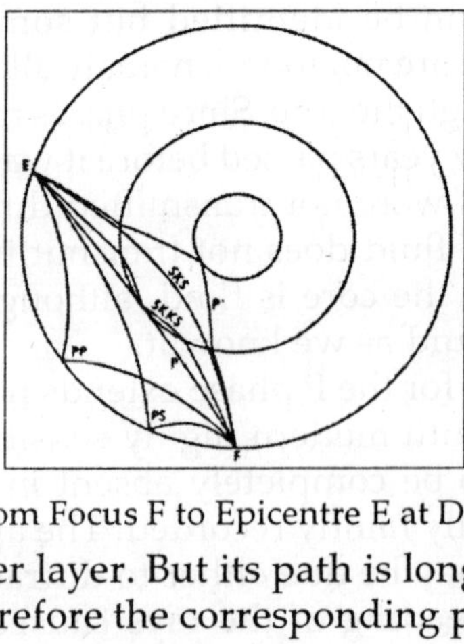

Fig. Rays from Focus F to Epicentre E at Distance 108°.

Faster in the lower layer. But its path is longer, and energy is lost on refraction, and therefore the corresponding phase in a seismogram will be smaller than that due to the direct wave. There therefore appears a small P phase succeeded by another much larger P phase. This was observed by Mohorovi Eiae, who gave the correct interpretation. He was interested in finding the depth of the discontinuity but did not have the means of determining it with any accuracy.

Actually it turned out to be extremely difficult to arrive at reliable values for the depth, though many different methods were employed. The best results have been obtained from explosions, which can be looked upon as artificial earthquakes. They can be timed with great precision, and the focus is exactly known; when they are well recorded at suitable distances more useful data are obtained than can be derived from earthquakes.

It now seems to be established that the discontinuity is at a depth of between 30 and 40 km under most continental areas. Under the deep oceans it is at a much smaller depth, only about 10-15 km under the water surface.

While the evidence for the existence of the Mohorovi Eiae discontinuity and of the other subdivisions of the Earth mentioned above is very clear, precise determination of their depths is difficult. It depends on very precise determination of the velocity variation throughout the Earth, and this in turn depends on precise determination of the travel times of the direct P and S waves and of some of the reflected and refracted waves. Much important work has been done along these lines, a great deal of it in the 1930s. In the course of this work it appeared that subdivisions of the mantle have also to be considered, but the evidence for them is not of a very precise nature, and their location is uncertain.

It is a difficult and lengthy process to construct good time-distance tables or time-curves. To obtain accurate travel times we require to have the focus and the time of occurrence of the earthquake accurately determined. As a rule this cannot be done directly, because there are not enough observations close to and around the epicentre.

It is therefore necessary to make use of time-curves already in existence, and there is then the risk of transferring errors from these curves to the new travel times. When time-curves have been determined from them it may therefore be desirable to have the elements of the earthquakes redetermined and the whole process repeated.

For the construction of the time-curves it has been customary to use a graphical method, plotting travel times against distance and drawing a smooth curve through the cluster of points thus obtained.

It is a somewhat arbitrary process, and, in common with other smoothing procedures, it is apt to smooth away or faultily to introduce changes of slope or curvature.

This is serious, because it is such changes that indicate the existence of more or less abrupt changes of velocity or velocity gradient in the Earth. Many time-curves have been constructed in the course of time, and the existence of various so-called discontinuities has been derived from them.

Two sets of time-distance tables or time-curves are now available which are far better than any previous ones. They are due to Gutenberg and Richter and to Jeffreys and Bullen. In 1928 Gutenberg published his *Frankfurter Laufzeitkurven* based on a large number of observations. In collaboration with C. F. Richter he greatly extended the work. In 1936 they jointly published the first part of "On seismic Waves", containing time-curves for a great many phases.

Amplitude variation was considered for the fixing of the distances at which the curvature of the time-curve was either greater or smaller than usual. The amplitudes of the recorded waves should be relatively large at the distances where the time-curve bends strongly, and amplitudes should be small where the curve is straight. Amplitudes were measured and used, and although very precise information is not derivable in this way, some useful indications were obtained.

The readings of the seismic records from all over the world are collected and published in "The International Seismological Summary" for which foci and times of origin of indìvidual earthquakes are determined. For the reduction of the data down to 1928 inclusive the Zöppritz tables were used, but it became more and more apparent that the times given by these tables departed seriously from actual travel times.

In 1928 Jeffreys began his very important work on travel times by a preliminary revision of the Zöppritz tables, using I.S.S. data. Later he undertook a thorough revision in collaboration with K. E. Bullen. A great quantity of data was used, and for the first time statistical methods were applied and the accuracy of the results obtained was evaluated. This implied difficult and extensive work because of the complicated processes involved. In part, new methods had to be developed. After the main work many special investigations followed, improving somewhat the first results. In 1940 the Jeffreys- Bullen "Seismological Tables" were published, and these are now being used for the I.S.S.

From these tables Jeffreys calculated the variation of velocity with depth. On the whole, the velocities of P and S waves increase down to the core boundary. Jeffreys found the velocity of P waves to be 5.6 km/sec in the upper crust, below the sediments, and to be 13.6 km/sec at the core boundary, where it decreases to 8.1 km/sec. With increasing depth the velocity increases steadily in the outer core to about 10.4 km/sec. In the inner core it is nearly constant at 11.2 km/sec.

The velocity does not increase uniformly with depth all through the mantle. In the uppermost mantle the velocity increases slowly, but at a depth of a few hundred kilometres a strong velocity increase sets in, as indicated by a bending of the time-curve around 20° epicentral distance. Below a depth of about 1000 km the velocity again increases more slowly.

On the whole, the velocities as derived from the Jeffreys- Bullen tables are probably not far wrong, but there are serious uncertainties. This is chiefly because it is so difficult to determine the slope of the time-curve accurately. The boundaries of the mantle regions in which the velocity gradients differ are indicated by changes of slope and curvature of the time-curve, but our data are not accurate enough for us to say exactly where these occur. It has been found particularly difficult to determine the velocity variation near the boundary of the inner core.

A vast amount of data, in part more reliable, has accumulated since Jeffreys and Bullen constructed their tables. New seismographs have been developed, from the-records of which the arrival times of the phases can be read with greater precision.

In addition, explosion work has entered the picture. It has for a great many years been used for the exploration of the crust, especially for the finding of oil; since the second World War more effective explosives have been available, and many of the explosions have yielded results also for the upper mantle.

Some large accidental explosions have also provided seismologists with new data. It was then found that the velocity just below the crust was greater than that derived from the Jeffreys-Bullen tables. Jeffreys drew attention to this, and in later work he provided corrected P tables for distances up to 30° for Europe. For, to complicate matters, it turned out that there were regional differences not only in the crust but also in the uppermost mantle.

From the new tables it appears that the strong increase of velocity gradient in the mantle at first placed near 400 km depth is likely to occur at a much smaller depth, probably between 200 and 250 km. However, attempts to fix the depth accurately have as vet met with unsurmountable difficulties.

The study of the Earth's interior is approached from many different directions. The upper mantle plays an important part in many investigations, and geophysicists in various fields look to seismologists for precise and detailed information. Have we any means of supplying such information?

K. E. Bullen's presidential address to the International Association of Seismology and the Physics of the Interior of the Earth at Toronto in 1957 had the title "Seismology in our Atomic Age". He pointed out that atomic explosions had far greater energy than the chemical explosions it is possible to use, and that they send waves deep into the interior of the Earth. Some atomic explosions have been recorded by seismographs, but there has been a reluctance by the authorities concerned to give prior information about the exact location and time of the explosions.

Though this has reduced the value of these explosions for seismological purposes, a few important results have been obtained, results that seemingly could not be derived from earthquake observations. There is no doubt that if an opportunity arose to record atomic explosions, by many seismographs placed at suitable distances from the source, information would be obtained that would help us to solve the problems that now confront us.

We are here, as Bullen said, in a tantalizing position, for the tools we so much need exist, but they are not very likely to be placed in our hands. With ever-increasing fear of the perilous effect of atomic explosions, seismology can scarcely hope for any to be organized for its special purposes. However, if the intention to explode bombs for other purposes is made known beforehand and the location and time are communicated, as has been the case in some recent instances, it should be possible for seismologists to make some use of them.

It is also well to remember that great masses of earthquake data are as yet unreduced and that, skilfully handled, they may yield fruitful results. New methods are also forthcoming in earthquake studies. The surface waves, not dealt with here, spread over the surface of the Earth, but they penetrate to some depth below it; in large earthquakes they may penetrate to considerable depth. The intense study of surface waves carried out in recent years, especially at the Lamont Geological Observatory, has provided a new approach to the exploration of the crust and the upper mantle.

While seismology teaches us a great deal about the interior of the Earth there is certainly very much more we should like to know. If we ask what are the materials inside the Earth, in what state they are, what is their density, and so on, seismology alone does not supply the answer. We have to look to other branches of geophysics and to other sciences such as geology, the physics and chemistry of the Earth's interior, and also to astronomy for additional information. Much attention has been given to pertinent questions in recent years, but the results are, in part, highly controversial.

Important results on the density variation throughout the Earth were obtained by K. E. Bullen. The velocities of the seismic P and S waves depend on the density of the transmitting material and on the elasticity as characterized by the rigidity and the incompressibility. These three quantities cannot be derived from the two velocities, but estimates can be arrived at when information from various other sources is taken into account.

Bullen finds that the density increases in the mantle from about 3.3g/cm^3 just below the Mohorovi Eiae discontinuity to about 5.5 g/cm^3 at the bottom of the mantle It then jumps to about 9.5 g/cm3 and increases to 11.5 g/cm^3 at the bottom of the outer core. In the inner core there is a strong increase. The rigidity that represents the resistance to shearing stress increases in the mantle, until at the bottom it is nearly four times that of ordinary, steel. In the outer core the rigidity is quite small, and this is what we mean when we say that the core is fluid.

On the other hand, the incompressibility or the resistance to pressure does not change materially at the boundary of the core. The pressure when evaluated was found to have reached about 11/3 million atmospheres at the bottom of the mantle and about 4 million atmospheres at the centre of the Earth. Following a long line of argument. Bullen arrives at the conclusion that the inner core is likely to be solid. The composition of the continental crust varies a great deal from one region to another. Granite is one of its main constituents, whereas this material is absent under the deep oceans, where the crust is much thinner. The upper mantle is believed to consist of ultrabasic rock rich in olivine.

A transition is likely to take place, perhaps from one form of olivine to another, in the region where the seismic velocities increase more strongly than elsewhere, this region beginning at a depth of a few hundred kilometres. It has for a long time been believed that the core consists of iron and nickel, but recent investigations have led to the conclusion that the outer core possibly consists of material not much different from that of the mantle but transformed under the prevailing high pressure. The inner core is still believed to consist chiefly of iron and nickel.

THE RADIOACTIVE EARTH

Without the heat from radioactivity it is probable that we would have had no atmosphere or oceans. Even if the ocean had existed, no land would have risen above it. Indeed, it is probable that the earth would have had a bare, rocky surface like the moon's, scorched by the sun in daytime and bitter cold at night.

But the story of the earth is a story of heat. Throughout earth history large amounts of energy have been continuously expended in mountain building, volcanism, and other activities which have formed the continents, oceans, and atmosphere. Except for the actions of the surface agencies, driven by heat from the sun, the energy comes from the interior of the earth and must have been at one time in the form of heat. To try to explain the occurrence of this thermal energy, we must consider two principal sources: the heat inherited from the formation of the earth and the heat generated in the breakdown of radioactive elements.

There are many arguments in favour of believing that the earth formed at a relatively low temperature. If this is true, a uniform distribution of the radioactive elements that we estimate were contained within the earth would have heated it sufficiently to have caused it to melt or partly melt. It is purely by chance that the sequence of events which we believe followed was such that the bulk of the earth stopped heating up again and remained fairly stable.

These purely chance events are as follows. First, if the mantle of the earth is solid and there is no convection (transfer of heat by movement of fluid material) in it, it must lose heat by the slow process of conduction in the upper regions. If there is convection, from melting or otherwise, the heat can be rapidly transported to the surface. Therefore, the temperature can never get very, much above that necessary for melting. In the lower regions heat may be carried out by radiative transfer.

Second, as soon as melting begins, there probably would be a migration of the molten radioactive substances upward because they crystallize at lower temperatures than compounds of magnesium, silicon, and iron. They would be forced upward in the liquid as the solid material settled downward.

If in any region the heat-producing elements have not moved close enough to the surface, the temperature will rise locally to the melting point and a further upward migration will occur. Eventually they will have come close enough to the surface so that there will be no further melting. Enough of the heat generated will be lost by conduction to the surface so that a stable solid mantle remains.

Gradually, following that time, the radioactive elements will decay slowly, and their heat production will diminish. This would tend to stabilize the mantle so that it is at some temperature below the melting point of its most fusible components. But if the chemistry of the radioactive elements had been otherwise and they had settled into the core, the earth would be continuously melting, losing heat by convection and solidifying again.

THE MAJOR HEAT-PRODUCING ELEMENTS

Let us now examine the amount of heat given off by radioactive elements and estimate what abundance of these elements would cause melting in the mantle. The element uranium breaks down through several stages to form a stable end product, lead.

As it undergoes successive transformations toward this stable end product, the isotope uranium 238 gives off 8 alpha particles as well as numerous gamma rays and beta particles. Summing up the energies of all these emitted particles and rays, we find that a total of 47.4 Mev (million electron volts) of energy has been expended for each atom of uranium 238 that breaks down to form an atom of lead 206.

Since 1 Mev is equivalent to 3.83×10^{-14} calories, it can be calculated that one gram of uranium in equilibrium with its daughter products is continuously giving off 0.71 calories per year. Similarly it can be calculated that the isotope uranium 235, of which atom bombs are made, is giving off 4.3 calories per gram per year when in equilibrium with its daughter products.

Thorium and its series give off 0.20 calories per gram of thorium per year. The only other important heat producing element is the isotope of potassium, K40. This gives off beta particles and gamma rays at a rate that yields 27×10^{-6} calories per gram of total potassium per year.

Average granite and volcanic rock contain approximately the following amounts of these radioactive elements:

Rock Type	Uranium parts per million	Thorium parts per million	Potassium%
Granitic rocks	4	14	3.5
Dark-colored volcanic rocks	0.6	2	1.0

Thus the radioactive components of the average granite can produce 7 microcalories of heat per gram per year. Other rocks that make up the bulk of the crust produce somewhat less heat than granites, and it is estimated that the average rock in the crust above the Mohorovi Eiae discontinuity probably produces about 2 microcalories of heat per gram per year. There is a measurable amount of heat continuously flowing to the surface of the earth. Measurements over continental areas have indicated that this amount averages about 1.2 microcalories per square centimeter per second. The amount of heat flowing in oceanic areas has been difficult to measure, but several measurements have been made. This is done by dropping a probe from a ship so it penetrates the mud on the bottom of the ocean for some distance.

Refined temperature devices are then used to record the difference in temperature between two points on the probe. By determining the thermal conductivity of the mud, it is possible to calculate the amount of heat flowing upward from the earth into the ocean water. Surprisingly, it turns out that approximately the same amount of heat is flowing from the interior of the earth in the oceanic areas as in the continental areas; namely, about 1.2 microcalories per square centimeter per second.

From all these figures it can be calculated that the average continental crust down to a depth of 35 kilometers produces about 1/2 microcalories per square centimeter per second from the radioactive breakdown of uranium, thorium, and potassium. This is about one half of the observed heat flow to the surface. It means that only about one half of the heat flowing to the surface comes from a depth of greater than 35 kilometers.

If you measure the amount of heat flow and estimate the thermal conductivity of the materials in the crust and below the crust, it is possible to estimate the increase in temperature with depth. The temperature-depth relationship. Note that the production of the heat in the crust greatly reduces the thermal gradient (rate of heat flow) at depth.

It follows that the temperature at depth is very much less than would be expected if one simply measured the temperature in deep mines or other openings in the earth near the surface and extrapolated this information to great depth. In fact, if there were no radioactivity in the crustal rocks, the observed temperature gradients at the surface would require that the mantle be molten at a shallow depth; this is not in agreement with the known geological stability of the region. The facts, therefore, support the hypothesis that the radioactive components of the earth are largely concentrated in the near-surface layers.

By calculating the temperature at which materials would be molten at depths of 100 or 200 kilometers, it is possible to estimate how much radioactivity must be in the near-surface rocks in order to keep the temperature gradient within known bounds. Attempts to do this have indicated that at least 0.2 part per million of uranium and 0.7 part per million of thorium on the average must be in the rocks down to 100 kilometers depth under oceanic regions. Since these amounts of radioactive elements would supply much of the observed heat flow to the surface, there must be little heat left flowing from the interior.

Thus we arrive at two important conclusions. The first is that very little of the original heat stored in the earth at the time of its formation is being lost, and, therefore, the earth is not cooling down at an appreciable rate, if at all. Secondly, we conclude that the major part of the earth's heat is coming from the breakdown of radioactive elements.

Since almost all the breakdowns occur within the near-surface regions of the earth, it is reasonable to infer that some process has moved the radioactive elements to this location from a presumably homogeneous distribution at the time of the earth's origin.

MIGRATIONS THROUGH THE MANTLE

Again we see the need for some process to have brought up from within the earth the materials that make up the oceans and atmosphere and the radioactive elements. In support of this requirement, we see that uranium, thorium, and potassium do in fact belong to the group of elements that form compounds of rather low stability and therefore would be most likely to move to the outer part of the mantle in any process in which partial melting was involved.

It is interesting that these conclusions do not violate the concept of an earth that is composed of materials similar to the iron and stone meteorites. The proportion of radioactive components in iron meteorites is very small indeed and would contribute a negligible amount to the heat of the earth if it were made of similar material.

The amount of radioactive components in stony meteorites has been measured carefully, and it is a striking coincidence that the amount corresponds very closely with that needed to give rise to the observed heat flow in the earth if it were uniformly of stony meteoritic composition. The fact that the radioactive components have migrated to the outer part of the mantle does not alter this interesting and supporting evidence.

Thus we see an earth in which the central part is rather slowly changing, if at all, in temperature and losing heat to the outside very slowly. Near the surface, however, there is an important balance in which the heat produced by radioactive elements can flow to the surface without causing melting unless the system is disturbed in some way. If at the margin of a continent the accumulation of sediments formed a low-conductivity blanket which also contained added amounts of heat-producing elements, this might be enough to cause melting at a depth of 100 or 200 kilometers and give rise to volcanic activity and other effects related to mountain building.

There has been much discussion and difference in opinion about the possibility of major convective overturns in the mantle down to the core boundary as a result of inhomogeneous distribution of heat sources. This process of convection could give rise to surface activity also and could be the cause of mountain-building events. In either case it is the heat from radioactivity that provides most of the energy for the dynamic events that have occurred at the earth's surface throughout geologic time.

THE AMSOC HOLE TO THE EARTH'S MANTLE

The earth sciences panel of the National Science Foundation (NSF) had just completed two days of hard work evaluating some 60 odd projects requesting research grants. The difficult job completed we relaxed to reflect on our work. Many of the suggested projects were excellent and most were of high caliber. Walter Munk remarked, however, that not a single one of them was of such a nature that a really major advance in Earth Science would result.

He suggested the panel invent a project which might strike directly at the roots of a major problem and forthwith suggested a hole to the Moho. If the writer deserves any credit for past or future association with this project, it is that he took Munk seriously and prevented momentarily adjournment of the panel. In the few minutes remaining he proposed referring the project to the American Miscellaneous Society for action. This was not a joke. It was done for several very good reasons.

AMSOC has no constitution, no officers, no members, only founders, many of whom are distinguished scientists. Here was a society which could act immediately — no need to wait for the next council meeting to have the proposal referred to a committee which would report to the council a year later. Gordon Lill was appointed chairman of the committee to proceed with the project and there followed the breakfast meeting in California mentioned by Bascom [1959] at which the specific plans were laid out.

NSF being unable to grant funds to a society without officers or a membership list made it necessary that AMSOC get a reputable sponsor. The National Academy of Sciences-National Research Council very kindly, and I might say, courageously, gave the AMSOC committee a respectable home. NSF funds for a feasibility study were then forthcoming and, once in hand, Bascom was appointed Executive

Secretary of the committee. The committee's history from this point on is one of record and can be left to historians.

The Meteorite Analogy and the Mantle

In 1850 Boisse proposed that the Earth's interior might be analogous in composition to meteorites. It was a brilliant proposal for its day and in a general way it is probably correct. In detail, perhaps, it is today being taken too seriously. It depends on how good the meteorite sample is, considering only those seen to fall, and how reliable a sample this is of the average composition of the body or bodies from which the meteorites were derived.

The great majority of stony meteorites are analogous to volcanic rocks such as crystal tuffs, lithic tuffs, and breccias. Are we only getting a sample of the outer surface of the meteorite parent? Olivine nodules from the Earth's interior where ejected from a volcano as bombs are found to be friable, the crystals loosely held together. This is probably due in part to sudden decompression and in part to the lack of a cementing matrix.

The mantle of the meteorite parent body might well behave in a similar manner upon being suddenly extracted from its environment. Were this the case the small particles resulting would be swept up by the Sun leaving only the somewhat more lithified stony meteorites and irons in orbits which might collide with the Earth.

The decompressed densities of the inner planets (and the Moon) are not the same. This must mean that their compositions are not exactly the same. It may merely indicate a difference in the degree of oxidation of the iron present. It could mean that the initial bodies forming the solar system varied somewhat in composition or that their initial compositions have been changed by some process which caused losses to surrounding space.

The uncertainties in the meteorite analogy cannot be resolved by further speculation. A hole or holes are required if we are to know something more definite about the chemistry of that 84 per cent of the volume of the Earth called the mantle. (The crust is less than one per cent of the Earth and we know quite a lot about it from observation. It seems quite reasonable to accept the meteorite analogy to the degree of postulating a Ni-Fe core.)

A half-ton sample of mantle would probably give more specific information than the 1400 odd meteorites now in collections. The meteorites could then be properly evaluated to give much additional information. It seems foolhardy to put an enormous effort into attempts to sample the moon or even the planets without finding out what is 5 km below the sea floor. The information from the hole is necessary to attack

Choice of Site

Drilling a hole in the Moho in a continent is not at present possible because it involves depths near 100,000 ft and temperatures too high for modern drilling equipment. The high temperatures also present difficulties in that many desirable measurements could not be made because the electronic devices and electrical cables could not withstand such conditions. Oceanic islands were considered during the early stages of the feasibility study. Aside from the fact that these are of volcanic origin, and hence one would be' drilling into the underpinnings of a volcano, the depth to be drilled would be too great to reach the mantle.

A small volcanic island rising from the ocean floor in depths of 15,000 to 18,000 ft would be about 40 miles wide at its base. Volcanic islands are in isostatic equilibrium. They represent loads in excess of the strength of the crust. If one assumes a density of 2.8 g/cc for the volcanic material this load would depress the Moho to 78,000 ft. Assuming the lowest reasonable density rather than the most probable one would give a depth in excess of 60,000 ft..

This leaves only one alternative, drilling a hole from a barge in the deep sea. The Moho can in many places be reached at depths of 30,000 to 35,000 ft below sea level and only about half of this need be drilled, but the overlying water presents some unique problems. The ship must either be anchored or maintained in position by some sort of automatic position keeping system.

In choosing a site the weather conditions must be considered. In general this led us to look for one at a latitute less than 25°. It must, if possible, be within 500 miles of a port where supplies and repairs can be obtained and exchange of personnel becomes feasible. The heat flow on the ocean floor must be less than two microcalories per cm^2 per sec. Assuming a conductivity of 0.005 cgs units, two microcalories gives a gradient of 40°C. per km or 200°C. approximately at the Moho which is somewhat too high for existing well logging instruments. Two areas seem to fulfill the above conditions.

One is about 120 miles north of San Juan, Puerto Rico, and the other south of Los Angeles from Guadalupe Island toward Clipperton Island. Seismic work is now under way to find places in these areas where a depth to the Moho is favorable and where the heat flow is comparatively low.

Chapter 2

The Earth's Broad Pattern

Since the dawn of civilization, men have been filled with curiosity about the earth on which they live. Why does a volcano erupt? What makes rain? What force causes the earthquake? What is the source of the water that bubbles up in a spring? As man's curiosity led him to seek solutions to such riddles, he often found that he was faced with new ones even more baffling.

How did sea shells become entombed in the rocks of high mountain ranges? Why does one stream have quicksand on one bank and solid rock on the other? What controls the beautiful geometric forms of snowflakes and other crystals? Why does one well yield water in abundance, whereas another dug to the same depth is dry? It would be interesting to know how early man attempted to solve such riddles. By comparison, we of today have signposts along the way, for reasoning man has built up the method of investigation and the compilation of knowledge that is known as *geology*—the science of the earth.

To some of earth's riddles, geologists have won the final answers. To others, the suggested answers are still tentative; and to still others, only the faintest glimmerings of light that may ultimately illuminate the way to final solutions have, so far, been discovered. Progress in geology has not been at a uniform rate. There have been periods, following fundamental discoveries, when outbursts of fruitful activity quickly revolutionized some of geology's theories and methods. At other periods there has been little advance.

At times geologists even followed the wrong trail, and progress in some branches of the science came to a dead end. Then more information and new skills were acquired until, finally, the accumulated dogmas were overthrown and a new start made. The first roots of geologic knowledge are lost in antiquity. The early Greeks and some of the peoples of other early civilizations made progress in geologic study, but their ideas were largely based on untested speculations, and little has survived. The modern science known as geology is of comparatively recent origin—the word itself is less than 200 years old.

Despite its youth, however, geology has already done much to stimulate and unshackle the thinking of mankind. The demonstration that sea shells and other fossils entombed in the rocks are but the remains of animals and plants that lived in the geologic past routed dogmas that had warped men's thinking for centuries. From the detailed study of the biological relationships of living and fossil organisms, coupled with geologic investigation of the sequence and changes of fossil assemblages with time, the doctrine of evolution emerged. This doctrine has profoundly influenced modern philosophical and scientific thought.

Evidence, well documented, that the landscapes about us are not static but are slowly changing, has not failed to stimulate the imagination of thinking men. The wheat farmer tilling the cold, wind-swept plains of Alberta is curious about the shells turned up by his plow, and becomes amazed when told that scientific comparison of these shells with living marine organisms shows that his farm was once the bottom of a warm, shallow sea.

Who can deny the thrill that comes with the realization that less than 20,000 years ago the site of Chicago lay under a sheet of ice such as enshrouds Antarctica today? Or that the green, well-watered hills of Scotland's Midland Valley were once the site of shifting sand dunes similar to those of the modern Sahara? Yet, preserved in the rocks and soils along Lake Michigan's shore and in sandstone quarries near the city of Glasgow are the proofs —as clearly recorded as are the deliberations of the Roman Senate preserved in the writings of Seneca and Cicero.

The science of geology has brought to mankind new conceptions of time, just as astronomy has revolutionized ideas of space. The rocks record events, some of which date back at least 1,800 million years, and throw into sharp perspective the short lapse of human history, as compared with that of the earth as a whole. There is a fascination in reading from the rocks the evidence on which we may reconstruct events of millions of years ago. It is this fascination that has led men to develop the science of geology—the attraction that will cause them to undertake deeper exploration of the earth's riddles.

MINERALS, WEALTH, AND POLITICS

Man's interest in the minerals and rocks of the earth's crust ceased long ago to be that of mere curiosity. There are sound practical reasons for his investigations. Our modern civilization makes many uses of the minerals and rocks that compose the earth's crust. Industry is almost wholly dependent on them. From minerals we obtain the iron, copper, aluminum, and other metals that make an industrial civilization possible. Our chief sources of power are the mineral fuels, coal and petroleum. In recent years, we have learned how to release stupendous amounts of energy from radioactive minerals.

Even our individual desires and needs are closely tied to the mineral industries. The bricks in our houses, the salt that seasons our food, the material that paves our highways, the gold and silver ornaments and precious stones with which we adorn ourselves—all have been won from mineral deposits in the earth's crust.

Man's avid search for the gold and silver, the copper and gem stones that pleased his vanity and brought him security and wealth began early in the annals of civilization. With possession of the minerals, he sought to refine and improve them and to discover new uses to which they could be put. As a result, the arts and crafts in metal and stone were born; and these, in turn, expanded into the vast industries we know today.

On the international scene, the power and wealth of a nation is largely determined by its supplies of useful minerals, its authority over the areas that contain them, and its skill in discovering and utilizing them.

In this age of political readjustment between nations, we know that the vast accumulation of petroleum in Iran and Arabia is a potent force in world politics. We shall be wiser in world affairs if we know how petroleum occurs, how it is discovered, and how its quantity may be estimated.

Without the economic urge to find and exploit the mineral wealth hidden in the earth, many of the great forward steps in geology would never have been made; for geology is the science of the mine and the quarry, of the oil field and the placer.

STUDY OF GEOLOGY

Although geology is a complex and varied subject, it is also a stimulating and interesting one. Relatively few of its problems are so simple that they can be solved directly by one method of approach. Many even require supplementary investigations with the techniques of other sciences. Geologists are constantly taking over from chemistry, biology, physics, and engineering new methods, data, and theories that can be adapted to their needs. Geologists, in turn, have contributed data and ideas to these bordering sciences. Progress in one science advances all the others.

Because of the complexity of its problems, geology has not advanced so far as has physics or mathematics. The geologist cannot move a volcano to the laboratory to observe the growth of its cone, nor can he spread a bed of coal on the laboratory table to watch for millions of years its development. Yet, these are the simpler phenomena of geology. Factors of size and time make experimental study of many geologic processes difficult and often impossible.

It is not possible to put a lava flow in a calorimeter and measure its output of energy. Faced with these apparently insurmountable difficulties, geologists have had to devise ingenious, indirect methods for getting the answers to many of their questions. Despite these inherent difficulties in subject matter, however, geologists have been outstandingly successful in predicting where to drill for oil or other mineral deposits, and in arriving at verifiable solutions of complex scientific problems.

Geologists, if they would be successful, must develop resourcefulness and imagination. They must be able to make sound decisions on the basis of incomplete, and even partially conflicting, data. In deciding where to drill an oil well or where to develop a gold placer, the geologist must, in many instances, evaluate and coordinate several kinds of evidence.

His fundamental guides, of course, are the data from geologic mapping and other geologic techniques. He may also need to consider results from geophysical exploration, data regarding production of other wells or placers, and miscellaneous additional evidence drawn from engineering, economics, chemistry, physics, and many other sources.

These very factors of complexity and diversity, together with the newness of the science, combine to make geology a vigorous, rapidly expanding field.

A student who selects geology as his profession has a wide choice of what he will learn and do. For his first two years of training, he will study more chemistry, physics, and engineering than geology. A sound elementary knowledge of these basic sciences is essential for many advanced geology courses.

The student will learn something of ordinary laboratory techniques, but will soon find that his main laboratory is not a building lined with bottle-filled shelves and machinery.

The geologist's laboratory is the bold cliffs of high mountain peaks, the walls of deep canyons, and the slopes of desert ranges. A part of his education will be spent in strenuous hiking and climbing in some mountainous area, perhaps far from civilization, where he will map the rocks and their structures and collect other geological data. Such "field work" is essential to geological training.

Upon completion of his training, the geologist will find many opportunities open to him. He may work for an oil company and travel the earth in search of new petroleum deposits.

He may direct exploration to find new bodies of ore in a mine. He may have the responsibility of estimating accurately the reserves of ore in the ground beneath a mining property or the amount of oil that can be recovered from a partially developed oil field. He may be called upon to decide which of several small mines is the best prospect for development and investment.

As an employee of a federal or state geological survey, the geologist may map rocks and mineral deposits, investigate conservation problems such as soil erosion and mineral depletion, or classify public lands as to their minerals, soils, water, and other natural resources.

Or the geologist may teach at a college or university, training future geologists, and at the same time, engaging in efforts to discover new principles or to unify and correlate old ones. Other opportunities for research are open in government work, in various research institutes, and in industrial laboratories.

In time of war, the geologist can serve by giving authentic information on problems of terrain, by discovering and assisting to develop critically short mineral supplies, and by selecting targets in enemy territory which, when demolished, will put an end to some vital industry of the enemy nation. He may sit at peace conferences and advise on the mineral resources of various nations and their resulting industrial potential.

There are also opportunities in commerce for the geologist. His geological knowledge can be used to good purpose in the development of a cement plant, or in the operation of a stone quarry, a brick yard, or a sand-and gravel pit.

Whatever path the geologist takes, it is likely to lead to widespread travel, for the whole earth is the field for his investigations.

THE BRANCHES OF GEOLOGY

Geology is such a large and varied field that only the briefest resume of its subject matter can be outlined here. In general, this book emphasizes *physical geology* (also called *dynamical geology*). Physical geology is concerned with the physical processes that operate on and within the earth—the processes that have given the rocks of the earth's crust their composition and structure, and the forces that have shaped the landscapes we see on its surface.

Many separate geologic sciences contribute to the broad field of physical geology. Among the more important are *mineralogy*, the science of minerals; *petrology*, the science of rocks; *geodesy*, which is concerned with measuring the form and size of the earth; *structural geology*, which seeks to interpret the structures to be seen in the rocks in terms of the dynamical forces that have produced them; and *geomorphology*, which deals with the origin of landscapes and with the changes that are constantly occurring in them.

Touched much more lightly in this book is the broad field of *historical geology*—the science that traces the evolution and development of the earth and of its animal and plant inhabitants with time. Historical geology is based, first of all, upon physical geology, but also draws extensively upon *paleontology*, the science that deals with the study of animals and plants of the geologic past, and on *stratigraphy*, the science that is concerned primarily with the order and sequence of the rocks that make up the earth's crust.

Another great field considered only briefly in this book is *economic geology*—the application of the science of geology to the uses of man, as illustrated in the finding and recovery of valuable mineral deposits from the rocks of the crust, the search for and recovery of oil and water from wells, or the study of the depletion of soils by erosion.

The subdivisions we have enumerated are not independent sciences. For example, physical geology could never have attained its present development without the concurrent progress in paleontology. In a broad view of geologic science practically every branch contributes in some measure to all the others.

THE EARTH'S BROAD PATTERN

Some of the answers may be gained through observations made with the aid of an ordinary plumb bob. The plumb bob is one of the oldest and most useful instruments of civilization. The early Egyptians used it in recovering land boundaries after the annual floods of the Nile.

Today, just as did our ancestors, we use plumb bobs in all surveying; in the construction of buildings, bridges, roads, and tunnels; in the making of all maps; and in measuring the size and shape of the earth.

THE EARTH'S GROSS SIZE AND SHAPE

The ancient Greeks noticed that the earth casts a round shadow upon the moon during eclipses. They also observed that the surface of any large body of water is curved, for only the upper part of a ship's mast is visible at a distance, and the ship appears to rise gradually out of the sea as it approaches. From these observations, the Greeks correctly inferred that the earth is, roughly, spherical. It is much easier for us to arrive at the same conclusion today. In modern airplanes, we can circle the globe in a few days. The curvature of the earth can be seen plainly in photographs taken from stratosphere balloons and rockets.

A few of the early Greeks made even more penetrating interpretations. More than 2,000 years ago, Eratosthenes, a Greek geometer and astronomer, first measured the curvature of the earth's surface. Then, assuming the earth to be spherical, he computed its dimensions. Though measuring techniques have been greatly refined, his reasoning is still used in modern geodesy.

Eratosthenes learned that in southern Egypt, at Syene, the sun shines vertically down a well only at noon on the longest day of the year. On such a day, he measured the angle between the plumb line and the edge of the shadow cast by the sun at noon in a well at Alexandria. Alexandria lies 5,000 stades (the ancient Egyptian stade equals about 600 feet) north of Syene. On the premises—

a.	That the sun is so distant that its rays to Syene and Alexandria are parallel;
b.	That Alexandria lies due north of Syene so that a plane through Alexandria, Syene, and the center of the earth also includes the noon sun;
c.	That the plumb line points directly toward the center of the earth; and
d.	That the earth is a sphere--

the angle between the plumb line and the shadow at Alexandria is equal to the arc of the earth's curvature between the two points. On these assumptions, the circumference of the earth is given by solving the equation:

$$\text{Circumference} = \frac{360^\circ}{\text{Angle of sun's rays to vertical at Alexandria}} \times 5000 \text{ stades}$$

Eratosthenes' earth was too large, but only 14 per cent larger than the figures now accepted. About a century later, Poseidonius the Greek philosopher, applied the same method to another are but was not so favored with compensating errors. The size of the earth deduced from his measurement was a quarter too small, and led to Columbus' error in mistaking America for India.

We shall see how, with refined modern measurements, the simple assumption that the earth is a sphere—which was satisfied by the early crude measurements—has had to be successively refined as succeeding measurements have become more and more accurate.

It would, indeed, be difficult to find a better example of scientific method, and of successive changes in theory made necessary by improved observations, than is afforded by the history of investigation of the figure of the earth.

Modern Measurements

In the Seventeenth and Eighteenth centuries, as the expansion of navigation and the accurate surveying of land boundaries became more important, Eratosthenes' method came into wider use. Arcs of meridian (North-South lines), equivalent in length to one degree of geographic latitude or to some simple fraction thereof, were measured at many different localities. It was found that a degree of latitude is longer near the poles than at the equator. Or, stated in another way, if we hold to the assumption that the earth is a sphere, its radius, as determined by observations near the pole, is notably greater than that deduced from observations near the equator. Here the dashed circle, with the centre P, illustrates the size of the earth based on observations made near the pole. The dotted circle, with the letter E at its centre, illustrates the size as determined from observations made near the equator.

Note the discrepancy in size. These measurements indicate that the earth is not a perfect sphere.

The inconsistency between the polar and equatorial measurements vanishes if we modify our assumption that the earth is a sphere and assume instead that it is slightly flattened at the poles—or, in technical words, that the earth is an ellipsoid.

The solid line is an ellipse with its centre at C. This line fits the data from both the polar and the equatorial observations.

Thus, the simplest model of the earth that conforms with modern measurements is an oblate ellipsoid—the solid figure that is obtained by revolving an ellipse about its shorter axis. In the ellipse indicated by a solid line, the short axis would be a straight line connecting the North Pole with the South Pole. In the drawing, the flattening at the poles is greatly exaggerated, for the earth actually does not depart greatly from a sphere. The measurements now accepted and used internationally as the basis for official mapping, are:

Equatorial Radius .	6,378,388 meters	(3,963.5 miles)
Polar radius	6,356,912 meters	(3,950.2 miles)
Difference	21,476 meters	(13.3 miles)

In determining the size of the earth, Eratosthenes did not concern himself with mountains and other irregularities of its surface.

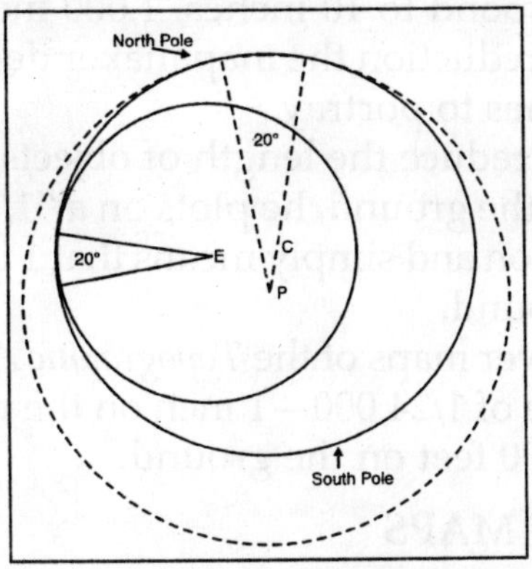

Fig. Circular Arcs Measured in Different Latitudes for the Earth

He regarded the earth as a smooth-surfaced sphere. Actually, this is a reasonably valid assumption, for the irregularities of the earth's surface in comparison to its size are not so great as those on the surface of a billiard ball. But to a Tibetan, living in the shadow of towering Mount Everest, such a comparison must seem a great over simplification.

How far does the earth's surface actually depart from perfect smoothness? To answer this question, we must find the difference in height between its highest and lowest points. This requires a kind of surveying different from that used by Eratosthenes, for now we are concerned with measuring the differences in altitude between various points on the earth's surface. The results of such surveys have been assembled in the form of *topographic* and *hydrographic maps*.

Maps and Mapping

Maps are the shorthand summary used by the student of the earth in presenting his data and observations. The earth's crust is complex. The intricate patterns of land and water, the forms of hills and valleys, the labyrinths that men have dug in mining are all so complicated in form that a true picture of them cannot be given by words alone. A map, however, condenses in intelligible form the findings regarding them.

Map Scales

A blueprint of a machine part or a dress pattern may be thought of as a map. Most of these are drawn as *"full-scale" maps. An inch on such a map corresponds with an inch on the object it represents.*

Few geographic and geologic maps, however, are full size. Most of them are *scale drawings*. In such "reduced-scale" maps, an inch on the map may correspond to 10 inches, 1,000 inches, 1,000,000 inches, or whatever unit of reduction the map maker deems desirable to show the features he wishes to portray.

If he decides to reduce the length of objects on the map to 1/10 of their true length on the ground, he plots on a "1/10 scale." The fraction is the ratio of reduction and simply means that 1 inch on the map equals 10 inches on the ground.

Many of the newer maps of the *Topographic Atlas of the United States* are drawn on a scale of 1/24,000—1 inch on the map corresponds with 24,000 inches or 2,000 feet on the ground.

LIMITATIONS OF MAPS

On a full-scale drawing, it is possible to show, for example, the head of a nail 1/10 inch across in full size. If the nail were to be correctly represented on a 1/200 scale, however, it would have to be drawn as only 1/200 of 1/10, or 1/2,000 of an inch across.

Such a point is too small to be visible. Thus, if such nails were to be shown at all on the 1/200 scale, they would have to be shown diagrammatically.

Their positions might be indicated, but their size must be greatly exaggerated if they are to be seen. This limitation of reduced-scale maps must constantly be remembered by the map user.

All maps are generalizations, drawn to perform a particular service or function. Therefore, all represent selections of data chosen to illustrate the particular purpose and are commonly exaggerated with respect to other data. A navigator's chart emphasizes the features useful to navigation; a good road map stresses highway junctions and, in doing so, may distort the distances between them.

The maps of Seattle Harbor illustrate one effect of map scale. Details such as the docks in Elliott Bay can be shown only on the larger scale map.

Whatever the scale, limitations in drawing and printing make it almost impossible for maps to be more accurate than 1/100 inch in the location of points. It is (difficult to make a legible pencil mark less than 1/100 inch across. A map of the United States reduced to a scale so that it could be printed on this page would be about 1/25,000,000; on it points less than 8 miles apart could not be shown as distinct without distortion. A thin line representing the Mississippi River on such a map would scale at least 4 miles in width.

Maps of the earth—the summaries of geographical knowledge—

have still another limitation: They must depict, on a *flat* surface, the *curved* surface of the earth. It is impossible to do this without distorting the distances between points or the angles between intersecting lines. Most maps are compromises between these evils. Some of the distortions that result from various methods of plotting.

TOPOGRAPHIC MAPS

The maps described thus far may be called *planimetric maps;* they show the relative positions of points but do not indicate their elevations above or below sea level.

The *relief* of an area is the difference in altitude between the highest and lowest points within it. Although, by means of skillful shading, a so-called *relief map* may give an impression of the relative steepness of the slopes in an area, such a map cannot be used to determine accurately the actual differences in elevation between any two points. It is impossible to read height accurately from such a map.

To meet this difficulty, geodesists have devised *topographic maps,* which are designed to show the elevations as well as the positions of points. They portray the three-dimensional form of the land surface, its *topography.*

A topographic map depicts a three-dimensional surface—one having length, breadth, and varying height above a reference plane or *datum* (usually *mean sea level*)—on a two-dimensional piece of paper. On such a map, lines called contours are drawn to portray the intersections of the ground surface with a series of horizontal planes at definite intervals above the datum plane. A convenient way of visualizing contours is to regard each of them as the shoreline if the level of the ocean should rise to exactly the height that the contour represents.

The 100-foot contour line on a map, therefore, shows the position of the shore if sea level should rise 100 feet. Every point on it is exactly 100 feet above the present sea level. Thus, each contour shows the course of a level line of definite elevation as it would appear if traced across the country shown on the map.

HYDROGRAPHIC MAPS

Hydrographic maps do for sea areas what topographic maps do for land areas. They not only depict the outlines of the water bodies but, by soundings (measurements of depth made by vessels at sea), they also show something of the topography of the bottom.

Although a few of the major irregularities of the ocean bottom became known during the Nineteenth Century, it was generally believed that the ocean floors were relatively much smoother than the surface of the land.

With the advent of sonic methods of sounding however, it soon became evident that at least a few parts of the ocean floors are highly irregular. The change in the interpretation of the bottom configuration of an area in the South Atlantic with the added data from sonic sounding.

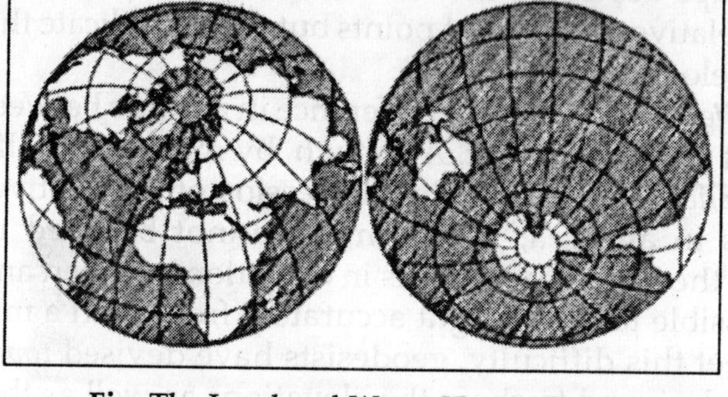

Fig. The Land and Water Hemispheres

IRREGULARITIES OF THE EARTH'S SURFACE

Land and Sea Areas

The major irregularities of the earth's surface are, of course, the continents and the ocean basins. Perhaps the most obvious fact is that most of the earth's surface is covered by the ocean. Using the best maps and charts available, the areas of land and sea have been carefully determined as follows:

Area of the sea	361,059,000 sq. km.	70.8%
Area of the land	148,892,000 sq. km.	29.2%
Total	509,951,000 sq. m.	100.0%

The land and sea are not uniformly distributed over the earth's surface. The surface of the earth can be divided into two hemispheres, one containing four-fifths of all the land, whereas the other is nearly nine-tenths covered by water. Let us now look at the topography of the earth's surface as it is recorded on maps of land and sea.

RELIEF FEATURES OF THE CONTINENTS

Relatively smooth *plains,* elevated and somewhat rougher surfaced *plateaus,* and rugged *mountain ranges* form the typical continental landscapes. Although details vary, several of the continents are characterized by lofty mountain ranges along one or more of their margins and by extensive plains in their interior. The highest and most extensive mountain systems on earth are in two great belts. One, the Circumpacific mountain system, is marginal to the Pacific Ocean; the other, the Alpine-Himalayan system, stretches across the southern part of the continent of Eurasia.

Surface Features of Eurasia

Eurasia, the largest continent, is also the most diversified topographically. Europe, the western part of this vast continent, can be divided roughly into three great topographic divisions:

- The northern highlands, which include rugged mountains rising 4,000 to 5,000 feet above sea level in the Scandinavian peninsula, and lower and more rounded ranges in the British Isles and northwestern France.
- The great central plain, which includes the vast plains of Russia, and which extends southwestward along the Baltic and North seas across Germany and into France.
- The Alpine system of mountain ranges and bordering highlands.

This includes the lofty ranges of southern Europe, such as the Alps, Pyrenees, Apennines, Carpathians, and Caucasus, and also extensive areas of lower highlands that form the typical landscapes of Spain, central and northeastern France, Germany, Czechoslovakia, and Yugoslavia.

The mean altitude of Europe is slightly less than 1,000 feet. Its highest point is Mount Elbrus, which rises 18,465 feet above sea level; its lowest point is the Caspian Sea, whose surface is 86 feet below sea level.

Asia is much more diversified than Europe. It has both the highest elevation and the lowest depression of any continent. Mount Everest in the Himalayas towers 29,002 feet above the sea. The Dead Sea lies 1,294 feet below sea level.

Asia is wrinkled by scores of lofty mountain ranges. Many of them meet in the Pamir knot of northwestern India, which is often called the "Roof of the World." Among the great mountain chains that radiate from the Pamir knot are the Himalayas, Kuenlun, Hindu Kush, Karakorum, and Altyn Tagh ranges.

Many other great mountain chains also diversify the face of Asia. Among the most prominent are the Elburz Mountains of Iran; the Tian Shan, Great Altai, and Khingan ranges, which stretch from west China northeastward into Mongolia and Siberia; and the prominent mountain chains that form the island arcs off the east coast of Asia, such as the

Kurile, Japanese, Philippine, and East Indies archipelagos. Between the great ranges of central Asia lie many high plateaus; among them are the Tibetan Plateau, the Tsaidan and Tarim desert basins, the great Plateau of Mongolia, and the Iranian Plateau.

Vast expanses of plain—the tundras and steppes of Asiatic Russia—occupy the area northwest of the central Asiatic plexus of mountain ranges. Other extensive plains form parts of China and India.

Surface Features of Africa

In general, Africa may be described as a vast plateau bordered in part by narrow coastal plains and diversified by numerous short, irregular mountain ranges. Unlike most of the continents, Africa has no long continuous mountain systems, although the Atlas chain, along the northwestern coast, can be followed with minor interruptions for nearly 1,500 miles. Most of this chain consists of low mountains, but a few peaks rise more than 13,000 feet above sea level.

Extending from Abyssinia southward to Lake Nyassa is a mixed group of irregular ranges and elevated plateaus diversified by two linear depressions —the *Rift Valleys* of east Africa. Lakes Tanganyika, Albert, and Edward lie in the western Rift Valley; the eastern Rift contains Lake Rudolf and part of Lake Victoria. The highest peaks in Africa are in this area.

The extreme southern and southwestern edge of Africa is bordered by the relatively low Cape Mountains.

Surface Features of North America

North America is mainly a broad central plain lying between marginal mountain systems. Toward the north this plain widens to include the plains of the Canadian prairie provinces and, on the northeast, the low rolling surface broken by inconspicuous mountains known as the Canadian Shield.

Far to the north, in the islands of the Arctic, Ellesmere Land and Grant Land, stand other mountains, higher than the Appalachians but little known.

The Appalachians, which border North America on the east, form a low but continuous series of small ranges and parallel ridges. The much wider and higher Cordilleran mountain system lies west of the central plain.

In the United States, the Cordilleran system is nearly 1,000 miles wide and includes the Rocky Mountains, the Cascade Sierra Nevada chain, and, near the Pacific, a series of lower and somewhat discontinuous Coast Ranges.

Between these mountain chains lie broad plateaus. The principal ones are the Colorado Plateau west of the southern Rockies, the Columbia Plateau between the Cascades and the Rockies, and the Interior Plateau of British Columbia between the Rockies and the Coast Range.

In Mexico, a high plateau, the Mesa Central, lies between two mountain ranges which merge into a high volcanic tableland south of Mexico City.

The highest point on the North American continent is Mount McKinley in Alaska, 20,300 feet, and the lowest is Death Valley in California, 276 feet below sea level. The mean elevation of the continent is approximately 1,300 feet.

Surface Features of South America

South America is similar in topography to North America. A long and high mountain system culminating in the Andes forms the continent's western margin.

On the east, from southern Brazil northward, a low and irregular series of highlands lies near the Atlantic Coast. The interior of the continent between these two mountain systems is a vast plain which, in Argentina, extends unbroken to the shores of the Atlantic.

The Andes are surpassed in height only by the Himalayas. Several of the Andean peaks rise more than 21,000 feet above the sea; the highest is the Argentine volcano, Aconcagua, 22,867 feet above sea level. At their northern end, the Andes split into three distinct chains. The easternmost extends into Venezuela and thence connects with the island arcs of the West Indies.

Between the parallel ranges of the Andes lie high intermontane plateaus. The Pacific coastline is bordered, especially in the southern part of the continent, by a low and rather discontinuous Coast Range.

Surface Features of Australia

Australia is a broad, low plateau. It is diversified on the east by the Dividing Mountains, a low series of ranges extending from northern Queensland to Tasmania. Other low ranges are found north of Adelaide and in parts of western Australia, but most of the continent has little contrast in elevation. The highest peak is Mount Kosciusko in southeastern Australia, which rises only 7,323 feet above the sea.

Surface Features of Antarctica

The continent of Antarctica is only partially explored. It consists of a great interior dome of ice rising to elevations of 6,000 to 10,000 feet, through which project peaks of some of the higher mountain ranges. Most of the continent appears to be an irregular plateau buried beneath glaciers. Mount Erebus, a volcano overlooking the Ross Sea, rises to a height of over 13,000 feet.

RELIEF FEATURES OF THE OCEAN FLOORS

The outstanding relief features of the ocean floor are portrayed. Just as the continents are diversified with mountains, plateaus, and plains, the sea floor also has its larger topographic features. Among these are the *continental shelves* and *slopes*.

The continental shelves are the submerged edge of the continents. Off most coastlines the bottom descends gradually from the shore to a depth of about 200 meters, or perhaps somewhat less, before breaking off abruptly to the much steeper continental slope.

The shallow continental shelves include more than 7 per cent of all the marine areas of the earth. They range in width from zero to as much as 800 miles (off the Siberian coast in the Arctic Ocean), and average about 30 miles in width for all coastlines. The surface of the continental shelves thus has an average seaward slope of about 20 feet per mile, though most of them are somewhat irregular and contain local ridges and depressions.

At depths that may be only a few meters or as many as 300 meters, the gently sloping surface of the continental shelf changes abruptly to the *very much steeper continental slope*. Off mountainous coasts, the continental slope may descend abruptly to the ocean floor at the rate of 300 feet or more per mile. Off wide coastal plains, its slope is generally about half this figure.

The continental slope off many shores is cut by so-called "submarine canyons". Some of these are deeper and more precipitous than the Grand Canyon of the Colorado, and form notches in the slopes recognizable to depths as great as 12,000 feet or more. Some head off the mouths of major streams such as the Congo and Hudson; others, equally impressive, head at the shoreline far from the mouth of any stream.

On the ocean floor itself, there is a diversity of topographic forms. Some are submarine mountains or other elevated tracts; some are depressions in the sea floor. Depressions may be grouped into the more-or-less rounded *basins*, the elongated *troughs* with gentle side slopes, and the similar but steeper-sided *trenches*. Any one of these depressions that has a floor more than 7,000 meters (23,000 feet) below sea level is called a *deep*.

Long, relatively narrow, elevated tracts are called *ridges*; or, if they are broader and larger, *rises*. Small, but significant submerged topographic eminences are the *sea mounts,* which are isolated steep-sided elevations; and the *banks,* which are shallower and broader, flat-topped elevations.

The Floor of the Atlantic Ocean

One of the greatest mountain ranges on earth, the *Mid-Atlantic Ridge,* almost bisects the Atlantic Basin from north to south. Its southern portion faithfully reflects the great bend of the coasts of South America and Africa. Throughout its length, the summit of the Mid-Atlantic Ridge rises approximately 10,000 feet or more above the floor of the Atlantic on either side. Most of its summit is less than 3,000 meters below sea level, and a few peaks—among them the Azores, Saint Paul's Rocks off the Brazilian coast, Ascension, Saint Helena, Tristan da Cunha, and Bouvet—project as islands above the surface of the sea.

The Walfisch Ridge, a range as high as the Alps and three times as long, branches from the Mid-Atlantic Ridge near its southern end and joins it to the west coast of Africa. The Rio Grande Rise, a lower and more gently sloping range, lies between the Mid-Atlantic Ridge and the coast of southern Brazil. In the North Atlantic, the Mid-Atlantic Ridge merges with the generally shallow bottom of the Norwegian Sea between Scandinavia and Greenland.

Two striking mountain arcs are present in the Atlantic. The Antillean Arc curves through the islands of the Lesser Antilles from the Venezuelan coast to the Florida shelf and separates the Caribbean Sea and the Gulf of Mexico from the Atlantic Basin proper. In the South Atlantic, the are of the Southern Antilles is, similarly, convex eastward.

It joins southern South America to the Antarctic continent through the South Georgia, South Hebrides, and South Sandwich islands. The only deeps in the Atlantic—the Puerto Rico Trough and the South Sandwich Trench, each more than 8,000 meters deep—lie on the convex sides of these arcs.

The Floor of the Indian Ocean

The Mid-Indian Rise, extending from India to Antarctica, resembles the Mid-Atlantic Ridge in lying medially in the ocean. It is broader and not so sharply marked off from the adjacent basins. Only about half its length reaches within 3,000 meters of the surface. West of the Mid-Indian Rise lie several other ridges, some of which branch from it.

Others extend southeastward from the African coast; Madagascar lies on one of these. South of Australia, the Indian and Pacific basins are separated by a moderate rise that joins Australia to Antarctica west of the Ross Sea. The area between Australia and southeastern Asia contains generally shallow seas, but the complex bottom topography of the East Indies includes several deep but relatively small basins and narrow trenches.

This area also contains many curving island arcs, the archipelagos of the East Indies. The Sunda Trench, south of Java, is the only deep in the Indian Ocean. Here, as in the Atlantic, the deep is on the convex side of an island arc.

The Floor of the Pacific Ocean

The Pacific Ocean floor is not well known. Apparently much of it is relatively flat. A sprawling, poorly defined rise extends from Antarctica to the coast of South America and across the southeastern Pacific to Central America.

A considerably shallower rise between New Zealand and Australia reaches far to the northwest off the east coast of Australia and the north coast of New Guinea.

This rise is notable for its steep eastern slope and for bordering several more-or-less isolated deep basins. East of the Tonga and Kermadec islands, it drops off abruptly to the Tonga Deep (more than 9,000 meters). Another great deep (9,400 meters), south of New Britain, is exceptional in that it lies on the concave side of the arc that connects New Britain and Bougainville islands.

The Hawaiian Islands are perched on a rather straight ridge, trending northwest. A steep-sided ridge almost at right angles to the Hawaiian Ridge has been discovered recently. It crosses the Hawaiian Ridge northwest of the Hawaiian Islands. Northwest of Hawaii, the ocean floor appears highly irregular, though chiefly deeper than 5,000 meters.

The most striking features of Pacific Ocean topography are the great island arcs that festoon its northern and western sides. Among them are the Aleutian, Kurile, Japanese, Ryukyu, and Philippine arcs. These island crowned arcuate ridges separate generally shallow seas—Bearing, Okhotsk, Japan, Yellow, and East China seas—from the Pacific Basin proper. Several comparable arcs branch southward from Japan through the Bonin and Marianas islands. These are not bordered by shallow seas; the Philippine Basin, on their concave side, is as deep as much of the main Pacific.

As in the Atlantic and Indian oceans, great deeps are common on the convex sides of the Pacific arcs. Among them are the Aleutian, Kurile, Japan, Bonin, Philippine, Ryukyu, and Marianas trenches. The only other known deeps in the Pacific Ocean are next to the Andean arc off the Chilean coast and between New Zealand and Antarctica.

The Floor of the Arctic Ocean

The Arctic is, of course, the least known ocean. In a sense, it is a land-locked mass of water, for the greatest depth connecting it with the Atlantic is only 1,500 meters. The Bering Strait, its connection with the Pacific, is only 55 meters deep. The Arctic is notable for the very wide continental shelves that border Siberia, Europe, and northern Canada. The sparse soundings that have been made in the Arctic have not yet revealed any great deeps. The greatest depth reported, 5,440 meters, north of Alaska, may even be in error, as indicated by later soundings in approximately the same area.

THE EARTH'S MAXIMUM RELIEF

Our brief review of the topography of the continents and ocean basins permits some generalizations about the earth's relief. It is in the great ocean deeps of the western Pacific that we find the maximum departure of the solid earth from sea level. Here, depths of more than 30,000 feet have been recorded by soundings. The greatest depth so far known is in the Philippine Deep, which descends to more than 34,000 feet, or approximately 6.5 miles below sea level.

The greatest height of land above sea level is Mount Everest, which rises 29,002 feet or nearly 5.5 miles. Great as these distances are by human standards, they dwindle into insignificance when considered in relation to the radius of the earth. If the largest circle that call be drawn on this page represents the earth, a moderate weight pencil line would, on the same scale, include within its width all the irregularities of the earth's surface, ranging from Mount Everest to the Philippine Deep. Thus, we see that, in relation to its size, the earth actually is about as smooth as a billiard ball.

From the best maps and other data available, careful estimates have been made of the total areas of the land surface that lie between various altitude limits. These are summarized in the accompanying graph. This graph brings out two important facts:

- The highest mountain peaks and the extreme abyssal depths of the sea are very small in area;
- Within the extreme range represented by the Philippine Deep and Mount Everest, the altitudes are not evenly distributed by area but there are two altitude ranges especially prominent.

One is the range between 4,000 and 5,000 meters (13,120 to 16,400 feet) below sea level, which embraces nearly one-fourth of the earth's surface. The other lies in the range from 200 meters below sea level to 500 meters above sea level and includes approximately one-fifth of the earth's surface.

The submerged portion of this second level, extending from a depth of 200 meters to the sea coast, is the continental shelf. At its outer edge, the sea floor descends rapidly in the continental slope to the ocean floors proper, which lie at the first altitude range mentioned, approximately 4,000 to 5,000 meters below sea level.

What are the reasons for these two dominant levels in the architecture of the earth? Why is the boundary between continental shelf and ocean floor so marked? These questions are among the great problems of geology. Their answers, so far as answers are possible, cannot be given in a sentence, but they are among the principal themes of this book. A part of the answer will be suggested by a more detailed discussion of certain aspects of the earth's relief.

Chapter 3

Gravity and Level Surface

From our study of the earth's relief, we found that mount everest rises approximately twelve miles above the level of the philippine deep. Although twelve miles may seem insignificant in comparison with the 4,000-mile radius of the earth, nevertheless the irregularities of the earth's surface pose some fascinating problems.

To one curious about the earth on which we live, several implications about the difference in level between its mountains and its ocean deeps are of lively interest.

An engineer knows that there are practical limits to the height of any building or tower. Even if we used the strongest rocks known, it would be quite impossible to construct a vertical tower as high as Mount Everest. The sheer weight of five and one-half miles of rock would crush the stone blocks at the base. Long before the tower could be built to such a height, it would come crashing down under its own weight. How, then, is Mount Everest supported at such a high elevation? Why does not the floor of the Pacific Ocean crush in and fill the Philippine Deep?

To gain some insight into how the great irregularities of the earth's surface are sustained, we must begin by studying the force that draws all objects toward the earth's centre. This force is *gravity,* one manifestation of the universal force of gravitation.

GRAVITY

Water in a pond and coffee in a cup both assume a *level surface.* If we hang a plumb bob above either liquid, we find *that the plumb line is exactly perpendicular to the liquid's surface.* Both the plumb bob and the liquid are responding to the force of gravity.

Each has approached as close to the centre of the earth's mass as its nature will permit—the liquid by flattening out so that it is as near the earth as possible, the plumb bob by bringing its mass as close to the earth as the string from which it hangs permits. The earth evidently has an attraction for each particle of the liquid, and for each particle of the plumb bob.

THE LAW OF GRAVITATION

The attraction of the earth for a plumb bob, for water, and for all other objects in the universe illustrates the universal *Law of Gravitation*, first formulated by the great English scientist Isaac Newton (1642-1727). This generalization is usually stated:

Every particle in the universe attracts every other particle with a force that is directly proportional to the product of their masses, and inversely proportional to the square of the distance between them.

The mathematical statement of this law is as follows:

F $\propto$ (read "varies as")

$$\frac{M' \times M''}{D^2}$$

In this expression, F is the force of attraction, M2 is the mass of one body, M3 that of a second body, and D the distance between the bodies.

As a numerical example of the way in which the law operates, consider the earth's attraction for a plumb bob that weighs one pound at sea level. At sea level, the distance of the plumb bob from the earth's centre is approximately 4,000 miles.

If this distance were doubled, the attractive force would be only one-fourth as great, for the denominator would be $8{,}000^2$ instead of $4{,}000^2$. In other words, if it were possible to suspend the plumb bob in space 8,000 miles from the earth's centre (4,000 miles above the earth's surface), it would weigh only 1/4 pound.

Gravity and a Level Surface

In making a topographic map, or in setting the floor joists of a building, what do we mean when we speak of a level surface? The surface of a pond is level, and it appears to be a plane but, as we know, in large bodies of water such surfaces are curved. A level surface is thus not a flat plane, but a surface that is, at every point, exactly perpendicular to the direction of the plumb line at that point. The direction of the plumb line is the vertical—it defines the *zenith*, the point directly overhead.

What Eratosthenes did in measuring the size of the earth was to determine the curvature of the level surface of the earth by measuring the angle between two such vertical lines; and then, knowing the distance between the vertical lines at the surface, he calculated the size of the hypothetical, perfect sphere he considered the level surface to bound.

There is another method, involving careful measurement of the exact force of gravity at various points on the earth's surface, by which we can determine some interesting things about the size and form of the earth.

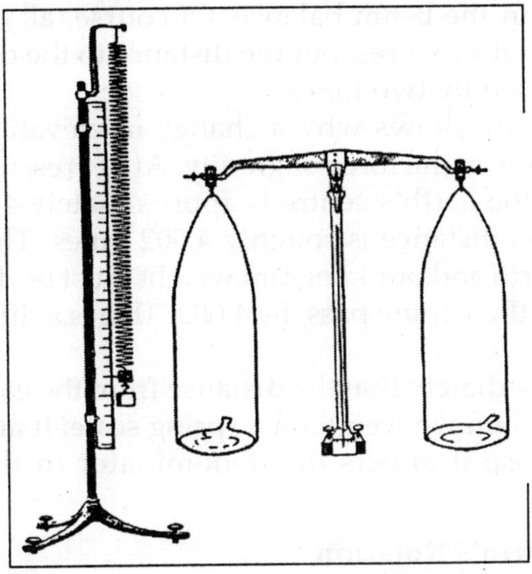

Fig. Two Common Weighing Instruments: Spring Scale

Weighing a Mass

When we measure the force of gravity, we are, in reality, weighing a mass, for the *weight of any mass is the measure of the force of attraction between the earth and itself*. There are two common methods of weighing—the spring scale and the beam balance.

A spring scale measures weight by the amount of stretching of its spring; it is calibrated by marking the stretching produced by the earth's pull on a series of standard weights. By international agreement, such calibrations are based on *the standard kilogram mass* preserved at the International Bureau of Weights and Measures at Sèvres, France.

At Sèvres, we calibrate our spring scale so that the pointer reads exactly one kilogram when the standard kilogram mass is placed on the scale. A second weight that brings the pointer to the same mark weighs, of course, exactly one kilogram and will counterpoise the standard kilogram mass when the two are placed in opposite pans of the beam balance. But, we shall find some interesting changes in these relations if we move our equipment from Sèvres to some other point on the earth's surface.

Effect of Altitude

If, for example, we take scale, weights, and balance to a pass in the Alps, two miles above sea level, we shall find that on the spring scale our kilogram weight will no longer weigh one kilogram but slightly less, though it will still exactly counterpoise the standard kilogram mass in the beam balance. Of course, all our equipment is identical with that at Sèvres, but the distance to the centre of the earth has been increased by two miles.

Newton's Law shows why a change in elevation above the sea results in a change in the force of gravity. At Sèvres, nearly at sea level, the distance to the earth's centre is approximately 4,000 miles; at the Alpine pass, this distance is roughly 4,002 miles. The product of the masses of the earth and our kilogram weight must be divided, at Sèvres, by 4,000 and, at the Alpine pass, by 4,002. This is a difference of nearly 0.1 per cent.

We see immediately that the distance from the earth's centre must be considered when we weigh on a spring scale. It affects the force of attraction because it affects the denominator in the expression of Newton's Law.

Effect of the Earth's Rotation

Factors other than altitude also affect the force of gravity on objects at the earth's surface. Because the earth rotates on its polar axis, there is a tendency for objects near the equator to be thrown off, just as mud is thrown from a spinning automobile wheel, or sparks from an emery grinder. The force causing this tendency is the *centrifugal force of rotation*.

It is very small compared to the force of gravity; hence, objects do not leave the equator and fly off into space. Nevertheless, the centrifugal force of rotation is not negligible and must be considered in our studies of the form of the earth.

As Newton showed long ago, this force should produce an equatorial bulge and a polar flattening in the figure of the earth.

Thus is explained the difference between the polar and equatorial diameters actually found by measuring arcs of meridian-Eratosthenes' method.

The force of gravity is greater at the poles than at the equator because the earth is not a perfect sphere; the surface at the poles is 13 miles closer to the earth's centre than at the equator. Even if the earth were a perfect sphere, an object at the equator would weigh less than the same object transported to one of the poles.

The centrifugal force of rotation is greatest at the equator and steadily diminishes to zero at the pole. The actual difference in the force that results from both of these factors is approximately 0.5 per cent. Thus, if a polar bear weighing 1,000 pounds in his natural habitat were transported to an equatorial zoo, he would there weigh only 995 pounds.

Effect of Local Variations in Rock Density

Still another factor produces variations in the weight of an object at different points on the earth's surface. This factor is the local distribution of rock masses of different density. Let us imagine that the gold buried at Fort Knox, Kentucky, as part of the United States Treasury reserve, formed a single block of pure metal. A cubic foot of gold weighs more than 1,200 pounds.

A moment's consideration of the inverse-square relation tells us that a sensitive device for weighing should show a greater weight for an object immediately on top of the gold than it would if the object were weighed on the roof of one of the great limestone caverns not far away, where much empty space is immediately beneath.

Thus, even though the latitude and altitude of two localities are identical, the force of gravity at each point must vary with the *density* (mass per unit volume) of the immediately underlying rocks. The rocks just below the earth's surface do vary considerably in density from place to place, though, of course, not to the extreme degree of our artificial example at Fort Knox.

THE GRAVITY PENDULUM

In summary, we can see that three factors—(1) altitude, (2) latitude, and (3) variations in density of the nearby rocks—affect the force of gravity at any point on the earth's surface. But the effect of each of these factors is small.

In order to learn much of value from studies of the variation in the force of gravity from one place to another we need a very sensitive instrument for detecting small variations in weight—far more sensitive than the crude spring scale of which we have spoken. Fortunately, we have precisely this kind of instrument in the *gravity pendulum*.

A free-swinging pendulum—that is, one which is not driven by clockwork or other external means—oscillates to and fro because of the force of gravity. If we suspend a heavy weight from a string, pull back the weight and release it, the weight falls toward the centre of the earth following the arc that the suspending string allows. It is being pulled along its path by the force of gravity.

But the *inertia*—the resistance that an object offers to any change in its motion—of the weight carries it past the lowest point on its swing, and causes it to rise against the force of gravity along the arc that the suspending string allows. It continues to rise until its inertia is counterbalanced and eventually overcome by gravity. Then it falls back again toward the lowest point, repeating the process again and again. The pendulum oscillates back and forth through the lowest point, rising in each oscillation to a little less height because of friction with the air. Gradually, the oscillations die down until, ultimately, the pendulum loses its motion and hangs straight down perpendicular to a level surface. It has become a plumb bob.

We borrow from physics the law that governs the *period of oscillation* (the time for one complete to-and-fro movement) of a pendulum:

The period varies inversely with the square root of the local acceleration of gravity, and directly with the square root of the length of the pendulum.

Newton demonstrated that this relationship explains why even good pendulum clocks show systematic gains or losses in time when moved about from place to place. We have seen that the *force of gravity* (the local acceleration of a freely falling body) varies from place to place on the earth's surface.

A pendulum clock that keeps good time in Paris would systematically lose time when taken to a point high in the Alps because the force of gravity (and hence the acceleration of the falling pendulum) is less at high altitudes than at low. From this effect of the force of gravity upon the period of a pendulum, we can measure gravity by means of a pendulum of known weight and length. To determine the force of gravity, we count the number of oscillations of the pendulum in a given time and calculate the force from the law for the period of oscillation.

Modern gravity pendulums are so constructed that they are almost frictionless. They are hung on knife-edge, jeweled bearings and are swung in chambers from which most of the air has been evacuated. The number of oscillations is counted by precision chronometers that can time the swing to less than 1/10,000 second. With such equipment, geodesists are able to measure the force of gravity with an accuracy of a few parts per million.

Ordinary gravity pendulums cannot be used on a ship at sea because of the rocking motion of the waves, but the Dutch geodesist, F. A. VeningMeinesz, has devised an adaptation of the gravity pendulum that can be used in a submarine submerged beneath the zone of strong waves.

Thus, in recent years, we have obtained thousands of measurements from the land and sea areas of the earth. From these measurements and from study of the angles between plumb lines at different points, geologists and geodesists have been able to make important deductions concerning the interrelations of gravity, topography, and the density of materials within the earth's crust.

Some of the most interesting of these interrelationships were discovered as a result of attempts to explain what appeared to be systematic errors in the location by triangulation of points on the earth's surface. Triangulation is the method of locating a third point by sighting on it from each of two points of known position.

But a point can also be located without recourse to triangulation by determining its latitude and longitude astronomically, as is done in navigation. Astronomic determinations of latitude and longitude are, of course, made with reference to the "vertical through the point," that is, the plumb line.

They are, in essence, determinations of the local zenith. If the plumb line invariably pointed directly to the earth's centre, determinations of position made by triangulation should coincide exactly with determinations made by astronomic methods. The plumb line, however, does not everywhere point directly to the earth's centre.

According to the law of gravitation, a plumb bob is attracted by a given mass one mile away with a force one hundred times greater than it is attracted by an equivalent mass ten miles away. We ought, then, to find the plumb line deflected toward a nearby mountain. This is, indeed, the fact.

Consider the situation in a deep Norwegian fjord. A narrow arm of the sea lies between massive cliffs 4,000 feet high. A plumb line suspended near one side of the fjord will be deflected toward the nearby mountain mass. Also, the sea's surface will be tilted slightly upward toward the mountain mass and downward toward the centre of the fjord.

Similarly, at the edge of a continent, the surface of the sea must be tilted upward slightly toward the continent by the attraction of the continental mass. Such tilts are very small, generally only a few seconds of arc, but they lead to appreciable errors in determining the position of a point by astronomic methods.

Also, from consideration of such distortions of the water surface, we see that the concept of an earth with the figure of an oblate ellipsoid, though a closer approximation to the truth than the concept of a sphere, does not precisely represent, in detail, the shape of the earth's levelsurface—that is, the surface which is everywhere at right angles to the plumb line.

Isostasy

Even the largest mountains, however, have very small masses

compared with the mass of the earth as a whole. Despite their being close at hand, we can expect their effect in deflecting the plumb line to be small.

To calculate the theoretical deviations operating at a given station by the law of gravitation requires considerable mathematical labour, for usually one has to calculate the effects of many irregular masses lying in different directions and at different distances from the observing station.

When this is done for a large series of stations, however, an exceptionally interesting relation appears, showing that mountain chains are not mere loads of rock heaped on the surface of the earth, but that they are sustained in some other way.

From such studies it has been discerned that mountain masses do not actually deflect the plumb line sideways as much as it would be deflected if the mountain were actually a load on top of the earth's crust. The same conclusion is reached from consideration of measurements of the force of gravity with the gravity pendulum. If a mountain were really a load heaped on the crust of the earth, the force of gravity (after correction for the effect of added altitude) should be greater on the mountain than on an adjacent plain because of the added gravitational pull of the mass of the mountain.

The results of many investigations with the gravity pendulum show, however, that over large areas there is little relationship between topography and the force of gravity, when this is corrected for elevation and masses above or below sea level.

From these relations, geologists and geodesists have inferred that the major irregularities of the earth's crust are sustained, not as loads borne up by the strength of the earth's crust, but by a process of flotation upon a heavier, plastic interior. Everyone knows that an oak plank will float lower ill water than a pine plank; it is denser and must sink deeper to displace its own weight of water. The idea is that *the excess mass of high areas,* such as mountain chains and continents, *is compensated for by a deficiency in density of the material of which the elevated tracts and their roots are composed.*

Mountain chains and continents float high because of their lower density. Plains and sea floors stand lower because either they are composed of rocks of higher density than the mountains, or else, if they are made of the same stuff as the mountains, this material is thicker under the mountains than under the adjacent lowlands. Such a condition of flotational equilibrium between blocks of the earth's crust is called *isostasy.* The word isostasy is also used to indicate a *tendency* for blocks of the earth to approach a condition of balance (flotational equilibrium) with adjacent blocks; that is, a tendency for heavy blocks to sink and light ones to rise, in the earth's plastic interior.

The theory of isostasy is an important concept that we shall refer to again and again. It will be well, therefore, to explain more fully the kind of geodetic evidence on which the theory is based by describing an actual example.

There is, perhaps, no better example than the first study of this kind ever made—the analysis of the deflections of the plumb line discovered in the course of the survey of the land mass of India. It was the investigation of these deflections that first led to the discovery of the condition of isostatic equilibrium in the earth's crust.

THE TRIGONOMETRICAL SURVEY OF INDIA

In the middle of the Nineteenth Century, the Trigonometrical Survey of India was organized under Sir George Everest (for whom Mount Everest is named) to carry out an extensive plan of precise surveying in order to establish "control points" for mapping that great subcontinent.

The method used for determining the relative position of the control points was triangulation. The triangulation was done very carefully and gave highly accurate distances between stations. Scores of points in India were located by this method, and their latitudes and longitudes were also carefully determined by astronomical observations.

If the latitude and longitude of one point in a triangulation net is known, it is, of course, a relatively simple matter to calculate the latitudes and longitudes of all the other points in the net with respect to the known one. We know the amount of curvature of the earth, and, hence, the distance along the meridian that will be subtended by an arc of one degree.

By the process of triangulation, we can determine the distance between two stations with a high degree of accuracy, and also determine the exact compass direction between the stations. It is, therefore, easy to translate measured distances, determined by triangulation, into terms of degrees of arc, and hence to calculate the latitude and longitude of each point in the net.

Such calculations are, of course, not dependent on astronomic observations at the new station. But if an astronomic determination of the latitude and longitude of the new station is made independently, it can be used as a check on the triangulation.

In the survey of India, it soon became apparent that, for some stations, determinations of their relative positions made by triangulation did not agree with those made by astronomic methods. At first, it was thought that perhaps errors had crept into the triangulation, but when checked the triangulation results were consistent. Also, several of the apparent "errors" were far too large to be accounted for by inaccuracies in surveying.

Two of the stations, Kaliana and Kalianpur, are among the stations discussed by Archdeacon Pratt, a British cleric who interested himself in the problem and who, in seeking an explanation for the apparent "errors," discovered the isostatic relationship. Now Kaliana is on the IndoGangetic plain, immediately beside the towering Himalayas.

Kalianpur lies far to the south, near the centre of the Indian peninsula. Archdeacon Pratt expected that, at Kaliana, the plumb line would be appreciably deflected northward toward the Himalayas, and, therefore, that the distance between the two stations, in terms of degrees of curvature of the earth's surface, would prove to be greater when calculated from the triangulation results than when based on astronomic determinations. This expectation was borne out.

Archdeacon Pratt was interested to find out whether the sideways gravitational pull of the Himalayas on the plumb bob accounted entirely for this discrepancy. Although no accurate topographic maps of the Himalayas were available, enough was known about the height and position of their principal peaks for Archdeacon Pratt to compute the approximate mass of the Himalayas above sea level and their average distance northward from each station.

Assuming that the mountains constitute an additional load on an otherwise uniform crust, he could then compute how much the plumb line should have been deflected northward at each station and the difference in astronomic latitude that should result from the deflection. He found that, on these assumptions, the plumb line at Kaliana should have been deflected 27.853 seconds to the north, and at Kalianpur it should have been deflected 11.968 seconds.

The difference between these is 15.885 seconds, over three times the 5.23 seconds difference actually found by the Trigonometrical Survey. The difference between Pratt's calculated results and the figure actually found was far too great to be explained by errors in triangulation, or by errors in Pratt's estimate of the volume and mass of the Himalayas.

Pratt's Theory of Isostasy

Pratt saw that some of his assumptions were not in conformity with fact. One of the assumptions implicitly accepted was that the density of material below sea level was identical, whether the material lay beneath the plain of India or beneath the Himalayas. He had assumed that the mass of the Himalayas, from their highest peaks to sea level, was a load heaped on a crust considered to be uniform in density below sea level.

Pratt saw that the discrepancy might be explained if the rock below the Himalayas were of less mass (lower density) than that to the same depth beneath an equivalent area of the plain of peninsular India.

He suggested that both the Himalayas and the peninsula are "floating" on a deeper layer of still denser earth material, and that the heights to which their surfaces rise above the surface of this deeper layer are inversely proportional to the average densities of the material composing the two "blocks".

In other words, the highstanding mass of the Himalayas is "compensated" or balanced by a corresponding deficiency of mass extending far below sea level beneath the Himalayas.

A simple illustration of Pratt's idea. In the upper diagram, blocks of four different metals, each of which weighs exactly the same amount and each of which has the same cross section, are shown floating in a pan of mercury, a liquid of very high density (13.6 gr./cm.^{3}).

The metal blocks have different lengths because their densities vary. The antimony block (density 6.6 gr./cm.^{3}) is longer than the lead block (density 11.4 gr./cm.^{3}) because its volume must be about twice that of the lead block in order to weigh the same amount.

The blocks sink in the fluid until they displace enough mercury to equal their weight. Therefore, the antimony and lead blocks will sink to the same depth in the mercury, but, since the antimony block has nearly twice the volume of the lead block, it will project much farther above the surface than the lead.

Similarly, zinc rises higher above the surface than iron, but all four metal blocks project into the mercury to the same depth—exactly deep enough to displace their weight of that fluid. Pratt assumed that mountains, plains, and ocean floors showed essentially the same relations.

Mountains are like the antimony block. They project higher above the surface of the "fluid" substratum because they are composed of material less dense than that which underlies a plain.

Pratt's suggestion was the first formulation of the theory of isostasy, and his scheme of "compensating" for differences in elevation above sea level by variation in density of the blocks below sea level is known today as the *Pratt Theory of Isostasy*. Pratt did not use the word isostasy; the condition of equilibrium which he discovered was named by Dutton some thirty-four years later.

Airy's Theory of Isostasy

The same volume of the *Transactions of the Royal Society* that contains Pratt's discussion also contains a brief contribution by G. B. Airy, the Astronomer Royal of Great Britain.

Airy accepted most of Pratt's reasoning, saying that this conclusion should have been anticipated because it could be shown that if masses as great as plateaus and high mountains were loads on a solid earth, no rocks could be strong enough to sustain them. The rocks beneath would break and flow out laterally until a state of balance was restored. The only possibility, then, is to regard mountain chains as floating masses, as Pratt did.

Airy, however, suggested a different mechanism of flotation. He saw no reason to believe that the density of the material immediately beneath a mountain is any different from that directly beneath a plain. If both blocks are of the same density, but of unequal thickness, the difference in surface elevation may still be readily explained. The thick block (mountain block) will float higher above the surface, but it will also sink deeper into the heavier "fluid" below.

The height of the mountain block is compensated by a "root" of light material that projects deeper into the underlying "fluid" layer and displaces it, just as an iceberg one hundred feet thick will float higher in water than one fifty feet thick, although it also projects deeper into the water. Airy's view is called the *Roots of Mountains Theory of Isostasy*.

This theory accounts for the "errors" in the India survey just as well as Pratt's theory, and accords much better with what we know and can infer about the composition of the materials beneath the surface. Therefore it is more widely accepted by geologists.

A simplified view of the "Roots of Mountains" theory. In the upper diagram, several blocks of copper, all with the same density but with different weights because they are of different lengths, are seen floating in a pan of mercury.

The longest blocks rise higher above the surface and also sink deeper into the mercury. The lower diagram shows how blocks of the earth's crust, all composed of the same kind of material, may, nevertheless, float to varying heights above sea level provided the blocks are of different thicknesses.

Good evidence will be developed to show that the earth's crust is not divided into simple blocks free to move past one another along frictionless boundaries. Also, the substratum is not a fluid, although it may respond by plastic flow to heavy loads of long duration and so act essentially like a very viscous fluid to these large stresses. We know, furthermore, that the strength of the earth's crust is considerable; loading must reach a certain amount before this strength is overcome and the rock beneath gives way and begins to flow.

The point to be emphasized here is that both Pratt and Airy concluded that the Himalayas do not stand at their present height because their huge pile is a load supported by the strength of the earth's crust, but because it is, in some way, sustained by flotation. The general validity of this idea also finds strong support in studies of the intensity of the force of gravity as determined from observations with the gravity pendulum.

THE GEOID AND THE SPHEROID

We have seen how the modern measurements of arcs of meridian at different latitudes forced the modification of Eratosthenes' "Figure of the Earth" from a sphere to an oblate ellipsoid.

But a moment's thought over the facts just discussed must show that still further refinement is necessary if we are to deal with the question in detail.

Though nearby mountain ranges do not deflect the plumb line as much as one might expect—a fact that we have already explained by the concept of isostasy—they nevertheless do deflect it. Thus, even the surface of the sea cannot be an oblate ellipsoid, at least near to the shores, for it is warped upward by the gravitational pull of the adjacent lands.

This warped surface of the sea, which is, of course, *everywhere at right angles to the plumb line, is called the* GEOID. It may be thought of as *the surface of the ocean, and as the surface to which the ocean would stand in a system of narrow sea-level canals cut through the continents.*

The geoid, or in other words the level surface corresponding to sea level over the whole earth, obviously does not correspond to any regular mathematical figure; unlike the ellipsoid, for example, it cannot be formed by rotating an ellipse on its axis. Hence, it offers great mathematical difficulties to geodesists who are primarily concerned with precise comparisons of measurements.

To get around this difficulty, geodesists have devised a theoretical figure of the earth which is called the *spheroid*. The spheroid is really the *world-average* form of the geoid. The spheroid will lie very slightly above the geoid in oceanic areas and somewhat below the geoid in continental areas. These relations are illustrated on a greatly exaggerated scale. When we recall that the oceans cover 71 per cent of the earth, it should not be surprising that the spheroid, or "world-average geoid," is only slightly different from the ellipsoid.

Gravity Measurements and Isostasy

To every point on the theoretical figure of the earth, the spheroid, there corresponds a theoretical value of the force of gravity which depends only on the latitude of the station. That is, the theory takes account of the centrifugal force of the earth's rotation. By swinging a pendulum it is possible to measure the actual force of gravity. It is clear that, in order to compare the measured and the theoretical values of gravity on a common basis, we must allow for several factors.

In the first place, nearly every continental station lies higher than sea level and hence is above the spheroid. It is farther from the earth's centre of mass than the theoretical point on the spheroid beneath. The effect of this is to make the measured value less than the theoretical. On the other hand, the material between the ground surface and the spheroid adds its local attraction to the measured value.

This should make it too great as compared with the theoretical value. Also, there are the disturbing effects of nearby topographic features. Obviously, many complex measurements and computations are necessary before the measured and theoretical values can be compared. When this work is done, however, as it has been for many thousands of stations on land and many hundreds at sea, a most significant fact is disclosed.

By convention the theoretical value is subtracted from the measured value as corrected for all these disturbing influences. The difference is called *the gravity anomaly*. Now, gravity anomalies that are computed after correcting for elevation of the station and for the mass between the station and sea level (called *Bouguer anomalies* after the French geodesist who first suggested this method) are chiefly negative on land areas and in general more highly negative the higher the station above sea. Conversely, the sea stations have slightly positive anomalies as a general rule.

What this means can be easily seen. When we allow for the gravitational pull of the masses between sea level and the station, our corrected value is too low. But in making this "correction" we have assumed *uniform density of material below the spheroid*. As the anomalies are in general greater the higher the station, it is clear that one of two factors has influenced our result: Either (1) the material *between the station and the spheroid* is not so dense and hence has not so great an attraction as we assumed it would have, or (2) the material *below the spheroid* is less dense beneath high land than beneath low.

But the material between the station and sea level cannot be far from the average rock in density. It must, then, be the second factor that has brought about our anomalies. But this is precisely the conclusion reached from plumb-line observations: *The excess mass of the rocks above sea level is compensated for by a deficiency of mass below sea level.*

Conversely, the positive anomalies at sea stations must be explained by the error involved in assuming that the oceans are underlain by material of the same density as that beneath the continents. If we replace this assumption with another—that the oceanic segments of the earth are underlain by *more dense material than the continents*—the anomalies are greatly reduced.

Thus, both the deflections of the plumb line and the relations between gravity measurements and the earth's relief indicate the same thing: The larger segments of the earth are roughly in floating equilibrium (isostatic balance), one with another. Large areas of highland stand above large areas of lowland because either: (1) They are composed of less dense materials than underlie the lowlands; or (2) if they are made of the same material, it is thicker beneath the highlands. We shall also show that, for some areas of limited size, isostatic equilibrium is far from complete—locally, indeed, there is good evidence indicating that masses of rock of the size of small mountain ranges may actually be held lower or higher than their equilibrium position by the strength of the crust.

The maximum size of the masses that can be so held out of their equilibrium position is still vigorously debated by geologists. But there must indeed be a maximum limit, for considerations of strength and scale compel us to conclude that the great earth blocks comprising the continents and the ocean basins, and even blocks of the size of the major mountain chains, are incapable of supporting themselves except essentially in flotational equilibrium, one with another.

Strength

The preceding discussion of isostasy assumes that the earth has a plastic core capable of buoying up the crust by flotation. Airy emphasized the point that a "solid" or rigid core, no matter how "strong" we think the hardest rocks are, could not be strong enough to support the earth's crust; therefore, the core must behave as though it were essentially a fluid mass.

What does this observation mean? A piece of rock we pick up is strong and does not behave at all like a fluid. We have also seen that the earth's surface is irregular in detail, with great mountain ranges towering thousands of feet above the sea and deep troughs submerged thousands of feet beneath it. The great mountain cliffs do not seem to be flattening out of their own weight. As far as we can see, they are made up of rigid and strong rocks capable of maintaining the topographic relief.

How can we reconcile this paradox between what we observe and the evidence that large blocks of the crust rest in flotational (isostatic) equilibrium upon a plastic interior? Must we conclude from these evidences of surface strength and of bodily weakness that the earth has a strong solid crust floating on an interior that is liquid? There are cogent reasons to doubt that the subcrust is really liquid. Perhaps a part of the answer may be found if we consider carefully just what is meant by "strength."

The strength of a body is defined as the force (load) per unit area that is required to deform that body permanently-that is, to break it or make it yield continuously. Thus, we say that a block of granite an inch square that breaks under a load of $30,000 pounds has a *compressive strength* of 30,000 pounds per square inch.

A steel cable with a cross section of one square inch that breaks when a weight of 150,000 pounds is suspended from its end has a *tensile strength* of 150,000 pounds per square inch. Each solid substance requires a definite force per square inch before it will rupture or flow.

Fluids have no strength. Water has no strength; it will flow into, and fill, all available spaces of a vessel into which it is poured. Tar has no strength; a piece of iron placed on it will eventually sink to the bottom. It is merely more viscous than water. Indeed, part of the technical definition of a fluid is that it yields continuously under the slightest load or stress.

Solids, too, can be made to flow—that is, yield continuously without rupture-under particular conditions of temperature and pressure. *But solids, unlike liquids, require that a definite threshold force per square inch must be applied before the strength of the material is overcome and continuous yielding begins.*

Effect of Temperature

Whether a particular solid will break or will flow under an applied force often depends on the temperature prevailing while the force is applied. At red heat, a bar of iron is still solid, but it will flow under relatively small stresses that would not cause it to flow at room temperature—a fact that is utilized in forging iron.

There is good reason to believe that many solid rocks buried deep within the earth's crust have yielded continuously by flowage in response to the heat and pressure of the earth's interior. The same rocks at the surface deform only by breaking.

Such flowage of solids should not be confused with liquid flow;

the rock does not melt, and a definite *stress* (equivalent to the strength of the material under the prevailing environment) must be applied before continuous yielding begins. The general rule is that *any particular solid is weaker at high temperature than at low.*

Effect of Size, or Scale

In everyday life, we seldom think of the effect of scale upon the strength of materials, although, as we observed earlier, engineers have to consider scale in providing strength for massive structures.

The earth is a large structure. How are we to think of rock strength in huge masses? Suppose we are to quarry a single block of granite the size of the state of Texas. It is roughly 1,200 kilometers across, and we want to hoist a piece one-quarter as thick as it is broad—that is, 300 kilometers thick.

Grant that we have a quarry crane capable of hoisting it. The rock is assumed to be flawless and to have the average crushing strength of granite, about 30,000 pounds per square inch. Will it be strong enough to allow itself to be hoisted without disintegrating? Obviously, it is impossible to test such a problem directly, but we can investigate the properties of such a block by using a scale model.

What would be the properties of a model reduced to a size suitable for our experiment? To obtain a convenient size for our model of the state of Texas, we could reduce the true length of 1,200 kilometers to 60 cm (about 2 feet). Our model would then be on a scale of 5/10 millionths of the original.

On this scale, the 300-kilometre thickness would be reduced in the model to 15 cm. (about 6 inches). The force of gravity does not concern us because both the original and the model are at the earth's surface. We can also use material of the same density as the original, about 3 gm/cm3. What other factors must be considered?

If the model is to act the same as the original—that is, if its mechanical behaviour is to be identical—there must be the same ratio of strength to size in the model as in the original. In order to obtain the same ratio of load in our model, we must use material whose strength is 5/10 millionths that of granite. The strength of the material in our model, then, will be

$$\frac{5}{10000000} \times 30000$$

Pounds per square inch, or .015 pounds per square inch.

In Hubbert's words:

It is difficult to envisage a solid of this weakness. A crushing strength of .015 pounds per square inch is the same as 1 gram per square cm, which, for the density of 3 grams per cubic cm, would be the pressure at the base of a column 1/3 of a cm high. Any column higher than this would collapse of its own weight. Yet the size of the reduced block would be such that its thickness would be about 15 cm (6 inches) and its total weight about 180 pounds. The pressure at its base would be about 45 grams per square cm or 45 times the crushing strength of the materials.

Consequently, if we tried to lift such a block in the manner indicated in the figure, the eyebolts would pull out; if we should support it on a pair of sawhorses, its middle would collapse; were we to place it on a horizontal table, its sides would fall off. In fact, to lift it at all would require the use of a scoop shovel. That this is not an unreasonable result can easily be verified by direct calculation upon the original block. For it, too, the pressure at its base would exceed the crushing strength of its assumed material by a factor of 45. The inescapable conclusion, therefore, is that the good State of Texas is utterly incapable of self-support!

Effect of Confining Pressure

The strength of a substance is increased when it is placed under pressure from all directions simultaneously—that is, when it is under *confining* (hydrostatic) *pressure*. Laboratory experiments show that the crushing strength of the limestone from Solenhofen, Germany, is six times as great under a confining (hydrostatic) pressure of 10,000 atmospheres as it is under ordinary laboratory conditions.

Its crushing strength under ordinary conditions is 25,000 pounds per square inch. Under a confining pressure equivalent to 10,000 atmospheres (147,000 pounds per square inch), its strength is 150,000 pounds per square inch. Such a pressure is equal to the weight of a column of granite about 22 miles high.

Presumably, then, it is also equal to the hydrostatic pressure at a depth of 22 miles within the earth.We see, therefore, that pressure due to gravity is a great factor in the earth's strength. In short, there is nothing really inconsistent in the apparent incongruity of an earth with "rigid and strong" rocks at its surface capable of maintaining relief features of considerable size yet so "weak" that it reacts almost like a liquid to large differential loads equivalent to the weight of a mountain chain.

Our chief difficulty lies in visualizing the way in which a substance must act in very large masses or under different pressure-temperature conditions from those familiar to us.

Chapter 4

Characteristic of Minerals

The Earth's crust is not a homogeneous substance. If we examine its surface during a walk along a stream or a drive along the highway, we seldom fail to find a variety of rocks and soils which differ in colour, in coherence, in density, and in other characteristics. If we pick up a fragment of rock or a handful of soil and examine it more closely, we find that it, too, is a mixture of different substances.

However, the individual particles that make up the rocks and soils are not mixtures.

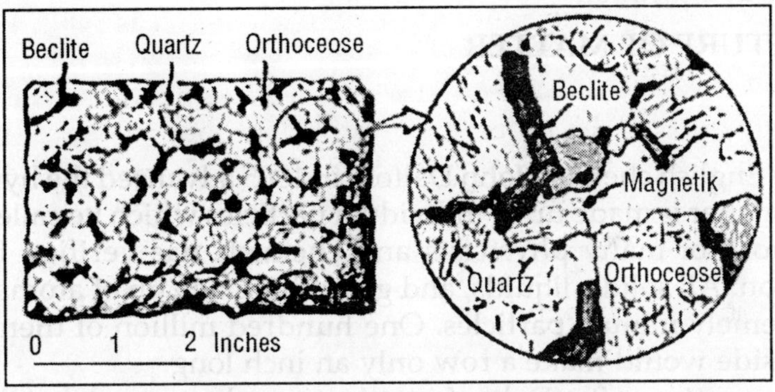

Fig. The Mineral Constituents of the Common Rock Granite.

Each is a distinct, homogeneous substance with definite chemical and physical characteristics. Some may be hard, transparent particles that resemble bits of broken glass; some may be dull, earthy grains; some may be tiny, elastic flakes that flash brilliantly in the sun.

Each of these distinct, homogeneous substances is a mineral. *Rocks and soils are aggregates of minerals*. Hence, if we are to understand the origin and classification of rocks, we must learn something about the various minerals that compose them.

A mineral is a naturally occurring substance with a characteristic internal structure, and with a chemical composition and physical properties that are either fixed or vary within a definite range. Minerals, then, are natural substances, found ready-made out-of-doors. The synthetic products that a chemist makes in a laboratory are not minerals, despite the corruption of the word in advertising. A druggist who tells you that a certain pharmaceutical preparation is "rich in vitamins and minerals" is using the term "mineral" in an entirely different way than it is used by a geologist. He is not prescribing a diet of mud and rocks.

To say that minerals have definite chemical and physical properties or properties that vary within certain definitely fixed limits is merely to point out that all particles of a single kind of mineral are alike in their physical and chemical characters, whether one particle comes from Brazil and another from the United States, or whether one may have crystallized in the shell of a snail and another from the water of a hillside spring.

The most definitive characteristic of a mineral is its internal structure, the core of our definition. To understand what we mean by internal structure, and, hence, the true nature of minerals, we must digress into a discussion of the fundamental structure and properties of matter itself. Here, the sciences of chemistry and physics unite with geology to arrive at a solution.

STRUCTURE OF MATTER

Atoms

The English chemist, John Dalton, in 1805 advanced the hypothesis that all matter is made up of individual particles which he called *atoms.* Later work of many physicists and chemists has verified Dalton's prediction. All solids, liquids, and gases are made up of atoms. Atoms are extremely minute particles. One hundred million of them placed side by side would make a row only an inch long.

Yet, a century after Dalton's work, it was discovered that the atom, small as it is, is composed of particles still smaller. Only three of the various subatomic particles that have been discovered are important in discussing the chemical behaviour of minerals—the *proton,* the *neutron,* and the *electron.* Every atom has a nucleus that is very small and dense. The nucleus contains one or more protons and, nearly always, one or more neutrons. There is an important difference between these two kinds of particles: *The proton has a unit of positive electrical charge, whereas the neutron is electrically neutral.*

The part of the atom that surrounds the nucleus is largely empty space, but within it the electrons of the atom revolve in orbits. Compared with the diameter of the nucleus, this outer part of the atom is very large—about 10,000 times as great.

Each electron has a unit of negative electrical charge. Each atom contains the same number of electrons as protons and, therefore, is electrically balanced or neutral. Electrons are very much lighter than protons or neutrons; they weigh only 1/1,845 times as much.

The orbits along which the electrons revolve about the nucleus are arranged in successive shells. The simplest atom, that of hydrogen, consists of a nucleus containing a single proton with a single electron revolving about it.

Helium has two protons and two neutrons in the nucleus, and it contains two electrons revolving in an electron shell close to the nucleus. In more complex atoms, the inner electron shell containing two electrons—the helium structure—is retained, but additional electrons, sufficient to match the number of protons contained in the nucleus of the atom, are added in one or more successive shells farther out from the nucleus.

The size of an atom, however, is not determined solely by the number of electron shells it contains. The calcium atom, which has 20 electrons distributed in 4 electron shells, is nearly identical in size to the sodium atom, which contains 11 electrons distributed in 3 shells. Both are very much smaller than the potassium atom which has 19 electrons distributed in 4 shells.

This matter of atomic diameter is highly significant in minerals one kind of atom can substitute for, or replace, another atom within the atomic latticework that makes up a crystal.

Elements and Compounds

From the chemist's point of view, the number of protons contained in the nucleus of an atom is fundamental, because the chemical characteristics of an atom are determined entirely by the number of positive electrical charges in its nucleus.

The number of these electrical charges, of course, equals the number of protons in the nucleus. All atoms having the same number of protons in the nucleus belong to one species of matter, called an *element,* or elementary substance. About ninety elements have been recognized in nature, and the atomic physicists and chemists have recently produced a few new ones by artificial means.

Each element has been assigned a definite *symbol,* such as H for hydrogen and Pb for lead—a convenient shorthand in writing chemical formulas and equations. The elements, together with their symbols, atomic numbers (number of protons in the nucleus), and atomic weights.

An elementary substance, such as hydrogen, oxygen, or iron, is made up entirely of atoms of a single kind, whereas a *chemical compound*, such as water, salt, or mica, is composed of two or more kinds of atoms bound together by electrical charges. We shall now consider the nature of these electrical bonds that hold diverse elements together in compounds.

For some reason, *those elements are most stable whose outermost electron shell contains 8 electrons*. Such elements—for illustration, the gases argon, neon, and xenon—are never found combined with other elements in the form of chemical compounds and, hence, are called the *inert gases*. The atoms of other elements may contain from 1 to 7 electrons in their outer shell.

The number of electrons in the outer shell is a major factor in determining the relative ease with which elements combine to form compounds. For example, the elements sodium (Na) and chlorine (Cl) are highly active chemically. Sodium has 1 electron in its outer shell, chlorine has 7. Sodium has only to lose an electron and chlorine only to gain one for both to attain the stable grouping of 8.

This is just what does happen when sodium combines with chlorine to form *sodium chloride* (NaCl), the substance we know as table salt. An electron is transferred from the sodium atom to the chlorine atom. But, the loss of an electron leaves the sodium atom no longer electrically neutral; it now has one more positively charged proton in the nucleus than it has negatively charged electrons in its electron shells.

Hence, the atom has one unbalanced positive charge. Similarly, the chlorine atom, by gaining an electron, acquires one unbalanced negative charge; there is one more electron in its electron shells than there are protons in its nucleus. Such a charged atom, with the number of its protons either more or less than the number of its electrons, is called an *ion*.

As unlike charges of electricity attract, and like charges repel, if the positively charged sodium ion is free to move, as in a solution or a gas, it is drawn to the negatively charged chlorine ion and the two may be held together to form a molecule of sodium chloride. *Molecules* are distinct groups of two or more atoms tightly bound together. The compound, sodium chloride, has very different properties from either of the two elements that combined to form it.

Two atoms also combine to form stable outer shells of 8 electrons without actual transference of one electron to another. This happens by the "sharing" of one or more electrons in such a way that, if the shared electrons are counted as belonging to each of the bonded atoms, each atom achieves a stable shell of 8.

Thus, two chlorine atoms with 7 electrons in their outer shells may unite to form a chlorine molecule by sharing two electrons, as shown in the following structural diagram:

:C̈l.	+	:C̈l.	→	:C̈l:C̈l:
Chlorine atom		Chlorine atom		Chlorine Molecule

The chemist's concept of molecules thus helps to explain the formation of chemical compounds and also the structure of gases and liquids in which most chemical reactions take place. The molecular concept of matter is also useful in describing the structure of some solids.

With most solids, however, a somewhat different view of the structure of compounds, a view in which the idea of molecules is not strictly applicable, seems to be needed. Most minerals are solids of this kind, as has been proved by many studies within the past twenty-five years. Some of these studies throw much light on the question of how elements combine, as well as on the internal structure of minerals.

STRUCTURE OF CRYSTALS

Geometrical Form

Everyone is familiar with crystals of certain minerals: halite (or rock salt), garnet, quartz (also called rock crystal), ice. Some crystals occur in beautiful geometric forms with surfaces bounded by smooth planes called *crystal faces*. Even the early Greeks noticed that some minerals, such as garnet and quartz, commonly have characteristic crystal forms.

Perfect crystals are rare. Most snowflakes fall as beautiful, six-sided, perfect crystals; but frost on a windowpane shows much less perfect ones, and the granules of ice that form on the surface of a freezing pond may show few, if any, crystal faces. This is also true for other minerals. Relatively few are completely bounded by plane faces and some show none at all.

Nevertheless, study of the common imperfect crystals, together with the relatively rare perfect ones, permitted mineralogists (as geologists who specialize in minerals are called) to make sound deductions about the internal structure of minerals and the fundamental properties of matter long before physicists and chemists had proved the atomic theory.

The first important step in the analysis of minerals by their crystal faces was made by Nicolas Steno, a Danish physician who lived in Florence, Italy. Steno, one of the outstanding figures in the history of geology, also made fundamental contributions to our knowledge of the origin and structural relations of rock strata.

Constancy of Interfacial Angles

Steno showed, with the crude instruments at his disposal, that, if crystal faces are found on a specimen of quartz, they always meet in characteristic angles regardless of the size and gross shape of the crystal.

An Italian student, Guglielmini, showed in 1688 and 1705 that similar relations held for other minerals. He also noted that the angles characteristic of one species of mineral differed from those of another.

Thus, in halite (NaCl) the angle between adjacent surfaces is always a right angle; this means the crystal may be a cube, a rectangular, boxlike figure, or any other right-angled parallelepiped. In quartz, as Steno had found, the angles between the long crystal faces that form the sides of the crystal are always 120°.

Steno's and Guglielmini's methods were refined and extended by later workers until thousands of similar measurements on many kinds of mineral crystals had been made. From them, mineralogists long ago reached the conclusion that the internal structure of each kind of mineral is unique.

They reasoned that the constancy of the interfacial angles in different specimens of the same mineral, regardless of the size and shape of the crystals, could mean only that each mineral is built up of minute particles regularly packed together in a definite geometric pattern. The pattern of packing determines the angles between faces and is identical in all specimens of a particular mineral. The size of the specimen depends merely on the number of such particles it contains.

Optical Properties

Other studies of minerals from a very different viewpoint fortified this conclusion. Among the most important were the studies of the effects that crystals produce on light transmitted through them. The Dutch physicist, Christian Huygens discovered the phenomenon of "double refraction" while studying the transmission of light through the mineral calcite.

One can easily observe this phenomenon by placing a fragment of transparent calcite above a dot on a sheet of paper. Instead of one dot, two are seen. If the calcite is revolved slowly, one dot describes a circle about the other. More than 100 years later, physicists showed that this phenomenon could be explained in terms of a theory of light.

According to this theory, the light ray which penetrates the crystal is broken into two rays which deviate slightly from each other as they travel through the crystal.

Furthermore, in contrast to ordinary light which vibrates in all directions at right angles to the line of propagation, each of the two rays that pass through the crystal vibrates in only one plane. Light so modified that it vibrates in only one plane is said to be *polarized.*

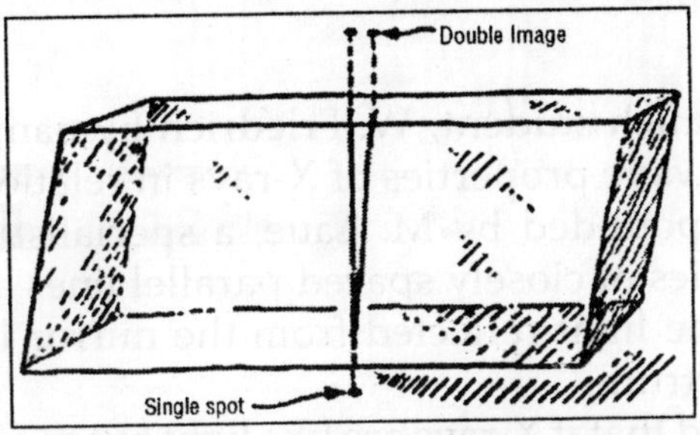

Fig. A Fragment of Clear Calcite Showing Double Refraction

These discoveries opened the way for an important new technique in studying and identifying minerals. William Nicol, who taught natural philosophy at Edinburgh, showed in 1829 that transparent fragments of calcite could be cut and glued together in such a way as to eliminate one of the two polarized rays. Others adapted these Nicol prisms to microscopes, making it possible to study the effects that crystals produce on transmitted polarized light.

Discussion of these effects is beyond the scope of this book, but it should be emphasized that the *optical properties* of minerals, determined by means of the *petrographic microscope,* are precise and diagnostic. By them, most minerals can be quickly identified. The petrographic microscope is the geologist's most efficient instrument for mineral and rock study.

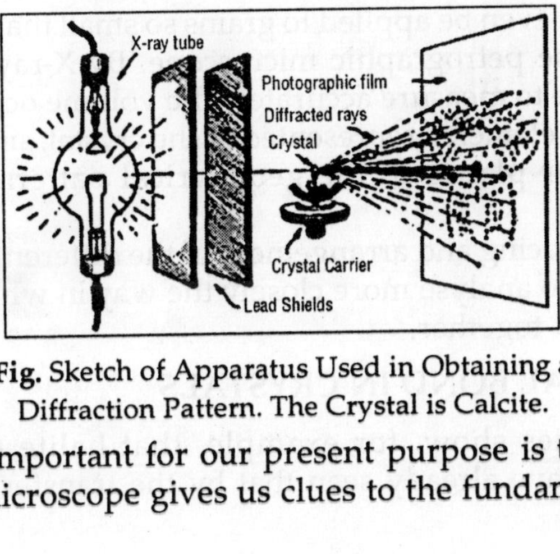

Fig. Sketch of Apparatus Used in Obtaining a Diffraction Pattern. The Crystal is Calcite.

But more important for our present purpose is the fact that the petrographic microscope gives us clues to the fundamental nature of matter and light. The behaviour of light in minerals is systematically related to the angles between the crystal faces. This suggests that light is influenced by very minute, systematically arranged particles within the crystal, and strengthens the conclusion from interfacial angles that minerals are made up of submicroscopic particles systematically packed together.

X-ray Studies

A young Munich student, W. Friedrich by name, was interested in the theory on wave properties of X-rays in relation to the structure of crystals, as expounded by M. Laue, a specialist in the physics of light. When a series of closely spaced parallel lines are scratched on a mirror surface, the light reflected from the mirror is broken into the colors of the spectrum.

Laue reasoned that if X-rays are like light but much shorter in wave length, and if crystals really are composed of submicroscopic particles geometrically packed in parallel planes, the surface of a crystal might act toward X-rays much as the ruled mirror surface does toward light.

In order to test this idea, Friedrich and a fellow student, P. Knipping, using a crystal of copper sulfate on an apparatus much like, developed the first "Laue X-radiogram." Later tests showed the same results as the first one, and gave definite and conclusive proof of patterns of crystal structure, confirming Laue's reasoning on the wave properties of X-rays. The experiments also proved beyond debate the inference of mineralogists regarding the internal structure of crystals. Also, a versatile new tool had been made available for the study of minerals.

By X-ray studies the geometrical arrangement of the atoms within a crystal, its *internal structure*, could now be worked out and The technique could even be applied to grains so small that they can hardly be seen with the petrographic microscope. By X-ray studies, it also became possible to measure accurately the volume occupied by atoms of almost all the elements represented in the crystal, since the distances between similar planes in the geometrical pattern could now be measured.

From the spacing and arrangement of the different kinds of atoms, it was possible to analyse more closely the way in which the atoms in crystals are held together.

THE CHEMICAL BOND IN CRYSTALS

X-ray studies show, for example, that halite (NaCl) has the structure. We have already seen that by the transference of the one lone electron in the outer shell of a sodium atom to the almost filled outer shell of a chlorine atom, both atoms can achieve the stable arrangement of 8 electrons in the outer shell. In acquiring this stable arrangement, each atom sacrifices its electrical neutrality and acquires a charge. It becomes an ion.

In a liquid or gas, two ions with unlike charges, such as a sodium ion and a chlorine ion, might be drawn together to form a molecule. In a crystal, the geometrical packing of the particles necessitates a fixed arrangement of the ions, an arrangement that must satisfy the electrical forces set up by the attraction of ions of unlike sign and the repulsion of those with the same sign.

Each positively charged sodium ion is equidistant from, and at the centre of, 6 symmetrically placed chlorine ions. Each negatively charged chlorine ion is similarly surrounded by 6 symmetrically placed sodium ions.

Therefore, we say that halite is an *ionic crystal* held together by the electric charges on its symmetrically arranged ions. Most minerals are held together by similar chemical bonds, although, in general, the internal structure is far more complex and not so easily visualized as in halite. Some minerals—diamond, for example are held together in another way.

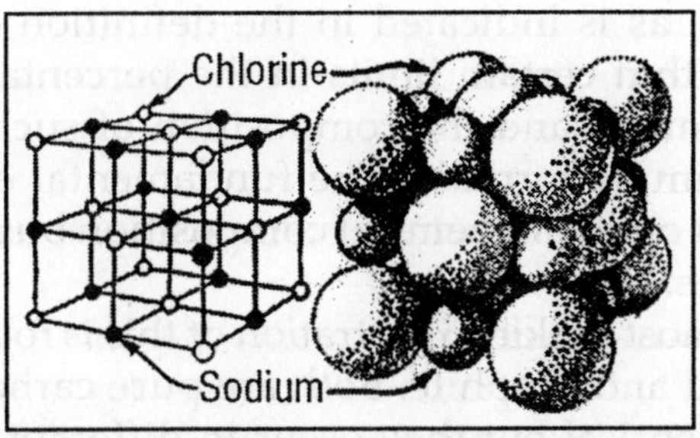

Fig. The Cubic Form, Right, and Internal Structure of Halite

Diamond is composed entirely of carbon; it is one of the crystalline forms of this element. Carbon atoms have 4 electrons in the outer shell. In crystals of diamond, each carbon atom is linked with four others.

This linking allows each of the 4 outer electrons in each carbon atom to be "shared" with an adjacent carbon atom. Thus, the carbon atoms in diamond achieve stability; each may be considered to have a complete outer shell of 8 electrons, though every electron is actually shared with a neighboring carbon atom. In terms of atomic structure, each electron may be thought of as oscillating between the orbits of the two neighboring carbon nuclei.

This close bonding of the atoms in diamond is very strong; hence, diamond is the hardest substance known. In this type of crystal, there are no ions—the atoms retain electrical neutrality but are selectively packed in positions so that electrons may be "shared."

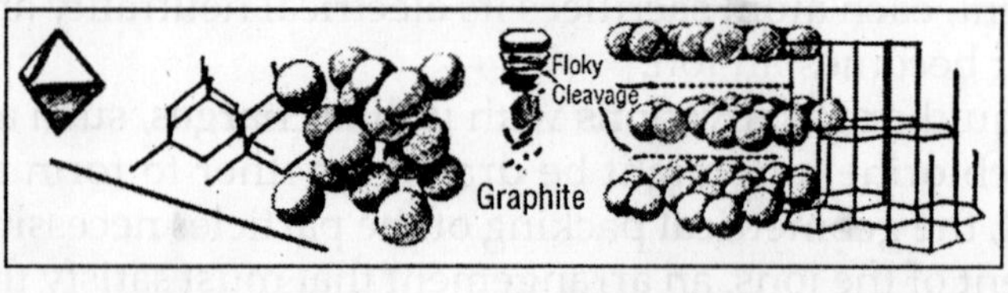

Fig. The Crystal Forms and Internal Structures of Diamond and Graphite.

It is readily seen that in either ionic crystals, such as halite, or in atomic crystals, such as diamond, there is no particular part of the crystal that can be considered a molecule. The chemist's concept of molecular association of atoms, applicable to gases and liquids, fails to apply to most minerals.

THE CHEMICAL COMPOSITION OF MINERALS

Some minerals, like diamond, sulfur, copper, and gold, are elements. Others, like ice (H_2O), quartz (SiO_2), calcite ($CaCO_3$), and kaolinite ($H_4Al_2Si_2O_9$), are compounds whose compositions can be expressed by simple chemical formulas.

Nevertheless, as is indicated in the definition of mineral, many minerals vary within certain limits in the percentage of the various elements they contain, and the composition of such a one cannot be expressed by a simple formula. The fundamental distinction among minerals rests not on their chemical composition but on their differing internal structures.

Perhaps the most striking illustration of this is found in the contrast between diamond and graphite. Both are pure carbon; their chemical composition is identical but they occur in different kinds of crystals. Diamond is the hardest known substance, graphite is soft and greasy. Most diamonds are transparent, graphite is opaque. Diamond is used as an abrasive and cutting tool, graphite as a lubricant because it yields fine flakes that glide readily over one another.

ISOMORPHISM

Certain elements may replace one another within a compound so that a range of chemical composition occurs within a single mineral species. Such replacement is called *isomorphism*.

The elements that substitute for one another may be chemically similar, but in many of the commonest minerals, the plagioclase feldspars, for example, they are not. Perhaps as simple an isomorphous series as any is the olivine group of minerals. The formula of this group is written $(Mg,Fe)_2SiO_4$, meaning that different specimens of olivine may have chemical compositions intermediate between the two "end members"; that is, they range from pure $Mg_2\ SiO_4$ to pure Fe_2SiO_4. It is only the iron (Fe) and the magnesium (Mg) that vary; the proportions of silicon and oxygen remain constant. The intermediate members of the series are regarded as *solid solutions* of the two end members because they are homogeneous crystals and not merely mixed aggregates of two minerals.

Mechanism of Substitution in Isomorphism

The application of X-ray studies to minerals has revealed much of the mechanism by which substitution of one element for another can take place in crystals. The controlling factor in isomorphism is not the number of electrons in the outer shell of the atoms of the two elements concerned, as might have been thought, but their atomic (or ionic) diameters.

These are measured in units of length called *Ångströms*. An Ångström is 0.00000001 centimeter long. In olivine, Fe and Mg can readily substitute one for the other, for not only does each contain 2 electrons in its outer shell, but the ionic diameters are very nearly the same.

In many mineral groups, sodium readily substitutes for calcium (2 electrons in the outer shell) because their ionic diameters are very similar (0.98 and 1.06 Ångströms, respectively). But sodium cannot substitute to nearly the same extent for potassium, although each has 1 electron in its outer shell, because the potassium ion is much larger (1.33 Ångströms) than the sodium ion.

This discovery was at variance with earlier chemical theory because, since sodium and potassium are very similar in chemical properties, it had been thought they would substitute readily for one another, as indeed they do in aqueous solutions. In crystals, however, a small amount of potassium may be substituted for sodium—the internal structure is warped to take care of the difference in ionic diameters—but, when the substitution exceeds a certain amount, the warping is evidently too great for the structure to remains stable; it breaks up into interlocking crystals of two distinct minerals.

Such warping of the internal structure during substitution, which shows the result of substituting calcium ions for about half of the iron and magnesium ions in the internal structure of olivine, giving the slightly different structure of monticellite, a mineral of closely related chemical composition.

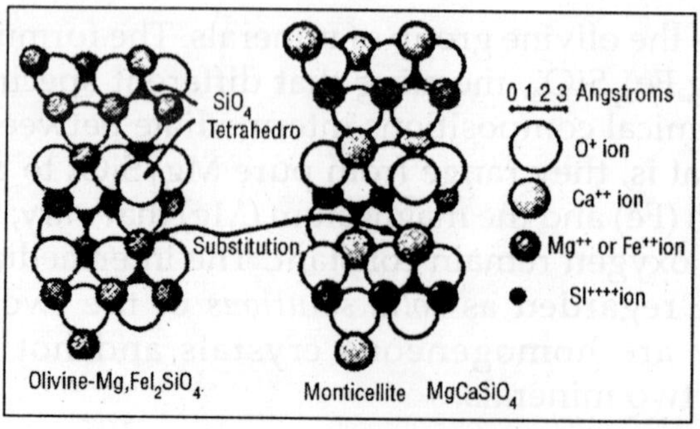

Fig. Structure of Olivine

Clearly, an element with one electron in its outer shell cannot be substituted for another with two electrons in its outer shell without destroying the electrical neutrality of the structure; therefore, a second, concurrent substitution is required to maintain neutrality.

For example, in the plagioclase series of solid solutions, the change from pure albite ($NaAlSi_3O_8$) to pure anorthite ($CaAl_2Si_2O_8$) takes place by the simultaneous substitution of Ca(2 electrons in outer shell) and Al(3 electrons) for Na(1 electron) and Si(4 electrons). Hence, chemical neutrality is attained, since $3 + 2 = 4 + 1$.

The ionic diameter of sodium is nearly the same as that of calcium; the ionic diameter of aluminum is sufficiently near to that of silicon so that the internal structure is not warped enough to become unstable. These slight changes in structure and composition, however, are sufficient to produce small variations in the optical properties.

In many minerals, similar substitutions of one element for another are numerous, and there may be several end members instead of only two. The composition of such minerals can seldom be expressed by a simple chemical formula in which the ratios of the elements are expressed in whole numbers.

Chapter 5

Classes of Rocks

Geology, like any science, has certain principles and generalizations that help to systematize and interpret the data collected by observation and experiment. The inquiring student should look critically into the validity of these generalizations. A basic principle in geology, often called the *Uniformitarian Principle,* was proposed by James Hutton of Edinburgh in 1785, and was popularized by the British geologist, Charles Lyell, in 1830. This principle of "uniformity in the order of nature" may be stated as follows:

"The present is the key to the past," or, applied more specifically to our present subject: Rocks formed long ago at the earth's surface may be understood and explained in accordance with processes now in operation.

The Uniformitarian Principle assumes the *uniform operation of physical laws throughout the geologic past.* It assumes, for example, that in the geologic past, just as today, water collected into streams and carried loads of mud and silt to the sea, that marine organisms lived and died in the ancient seas, and that their shells were buried in the sands and mud accumulating on the sea floor.

Hence, if features identical to those we now see in process of formation along streams and beaches can be recognized in and among the solid rocks, it is reasonable to infer that these features were produced by processes of the kind we now see in operation.

The Uniformitarian Principle, like any other scientific "law," rests on the circumstance that all the facts known conform to it. Long and careful searches have failed to find good evidence for ancient conditions totally unlike any existing today.

Yet, like most scientific laws, this one must be interpreted carefully and rather broadly. In applying the principle that "the present is the key to the past," we must keep in mind that, although there is good evidence to believe geologic processes have always operated in the same way, they have not necessarily always operated at their present rate or intensity.

Climates of the world were colder some 15,000 years ago and that glaciers were much more widespread than they are now; but there is every reason to believe that the glaciers of that time formed, moved, eroded, and deposited in precisely the same way that glaciers do today.

In the following discussion it will be apparent that some applications of the Uniformitarian Principle may be made to the problems of the origins of rocks.

THREE GREAT CLASSES OF ROCKS

The walls of many deep roadcuts show that the surface soil and the unconsolidated mantle rock beneath it form only a thin veneer. At depths of a few feet, we pass through them and enter rock. The deepest wells that have been drilled extend downward into the rocks about 20,000 feet—less than one-tenth of 1 per cent of the earth's radius. Materials now at greater depths cannot be examined, that the rocky shell extends to depths of at least several miles, and that some rocks now at the surface were once buried deeply.

Let us briefly summarize a few of the conclusions geologists have reached about rocks, leaving the evidence upon which these generalizations are based until later. Rocks are aggregates of minerals. Three great classes are recognized. The *sedimentary rocks* have been formed at the surface of the earth, either by the accumulation and cementation of fragments of rocks, minerals, and organisms, or as precipitates from sea water and other surface solutions.

The *igneous rocks* have formed from molten material that solidified on cooling. The *metamorphic rocks* have been formed by the transformation, *while in the solid state,* of pre-existing rocks beneath the earth's surface through the agencies of heat, pressure, and chemically active fluids.

Rocks are difficult to classify because they grade one into the other. Even the three broad classes of rocks intergrade, and not every rock can be placed unequivocally in one class or the other. Nevertheless, we can classify most rocks into these three great classes and also into smaller divisions, just as it is possible to classify people into tall and short, even though we know there is every gradation between.

Sedimentary Rocks

Everyone has observed how rills formed on a hillside during a downpour of rain spread sheets of mud, sand, and gravel at the base of steep slopes. Similarly, every stream, whether it be a small brook or a great river, carries unconsolidated debris downstream.

Most of the rock waste is dropped in sand bars or in beds of silt and gravel in the slack-water parts of the stream course, but it is carried further by later floods and most of it eventually reaches the ocean and accumulates in a delta at the stream mouth or is scattered over the beaches and sea floor by ocean waves and currents.

So commonplace and easily observed are the transportation and deposition of debris by running water that even the Greek philosophers learned to recognize water-borne deposits. They reasoned that beds of gravel and sand clinging to the walls of steep valleys high above the reach of presentday floods were the deposits of former streams. They saw in the clam and oyster shells protruding from some sandstone ledges on inland mountain ridges evidence of former higher-standing seas.

It was a more difficult step—and one probably not made by the ancients —to conclude that a hard, well-consolidated sandstone containing scattered fossil shells and exposed in the gorges of a mountain range is merely the *cemented*, shell-strewn sand of an ancient beach or sea floor that has been warped upward to its present height. In the Middle Ages, churchmen concluded that such fossils, many of which are not closely similar to the shells of animals now living in the sea, were not organic remains but "sports of nature," perhaps put in the rocks by the devil to confuse mankind.

Even the accurate and relatively enlightened German scholar, Georg Agricola, in whose words and woodcuts the late medieval Saxon mines and miners are still preserved for us, described only the leaves, wood, bones, and fish skeletons embedded in the rocks as organic remains. To him the fossil shells were "solidified accumulations from water" (whatever that may mean).

The restraints of tradition and authority were not thrown off until the geological pioneers of the late Seventeenth and Eighteenth centuries collected and compared the shells in hard rocks with those in unconsolidated sands near the seashore and, in the years between 1790 and 1815, made maps showing the surface distribution of strata of sandstone, limestone, and other rocks, demonstrating that soft sandstones may grade laterally into hard rock and that in some localities both types contain the same varieties of fossil shells.

Today, by modern tools such as the petrographic microscope, it is only the work of minutes to trace the steps whereby loose sand like that on the floor of the sea has been transformed to solid rock. It is easy to see that the individual grains of the fossiliferous sandstone, when magnified by the microscope, are of the same shapes and composed of essentially the same minerals as the sand grains of a modern beach.

The fossil shells of the sandstone, even though different from those of animals living today, show under the microscope structures so similar as to compel the belief that they are the remains of organisms that lived in the past.

Thus, the loose beach sand and the firm sandstone differ only in cementation. The voids or pores between the grains of the fossiliferous sandstone have been almost completely filled with mineral matter. By the filling of these voids, unconsolidated sand has been made into solid rock.

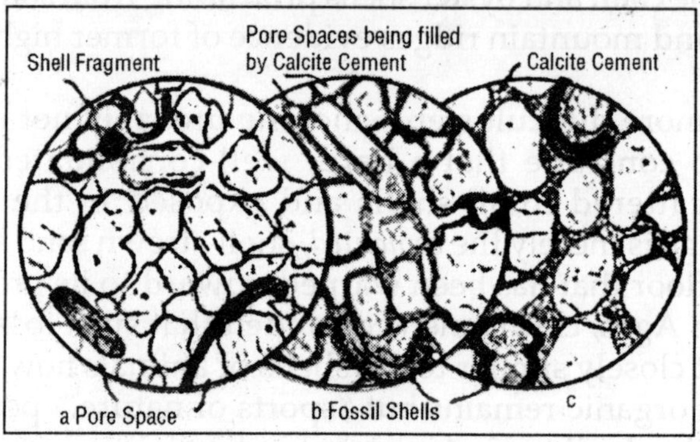

Fig. Cementation of Sand

The process of natural cementation is slow, but partially cemented rocks are common and, since all intermediate stages have been observed, the inference that sandstone is cemented sand is logical. By digging deeply into some modern beaches, we find that certain layers of sand below the surface, where they are not disturbed by wave action, are coated with a thin film of mineral substance that evidently has precipitated from the solution surrounding them.

Although this film may not have grown thick enough to bind the sand grains firmly, complete cementation will surely occur if the process keeps on for a sufficiently long time. More striking examples may be seen along certain stream beds in localities that have hot, dry summers. Here, the newly deposited gravel brought down by a spring flood may, under exceptional conditions, be completely cemented in a single season by the calcite and other minerals precipitated out of the water as the stream dwindles and finally disappears under the hot summer sun.

Thus, even modern sediments in process of accumulation show many stages in cementation from unconsolidated debris to solid rock. Yet cementation may also be extremely slow. Some ancient sedimentary rocks are only partially cemented and contain abundant open voids.

Streams, waves, and currents are not the only agents that deposit sediments. Glaciers and winds also move and deposit detritus which may then be cemented into rock. Organisms are also sedimentary rock builders. Off the shores of northern Australia and islands of the East Indies are great reefs many miles long, made up of the shells of corals, clams, marine snails, and a wide variety of other organisms which live in the shallow, clear water of these warm seas.

Even today, the living shells are being cemented together by lime-depositing plants and the interstices filled with microscopic shells and limy (calcite-rich) mud. Such accumulations of cemented shells form the common sedimentary rock called limestone. Limestones that show all the characteristics of these underwater reefs are also found on land far above the reach of present seas.

Even the New Guinea natives recognize that some of the limestone in hills near the shore has been formed in the same way as the reefs accumulating offshore today. Similar reef limestones, interstratified with other marine sedimentary rocks, are found in west Texas, the Balkans, and many other places thousands of feet above the sea all hundreds of miles from it. The conclusion that these rocks were once deposited in ancient seas as "coral reefs" of the kind that we now see in the southwest Pacific seems logically inescapable.

Coal is another organic sedimentary rock; the well-preserved cell structures and other plant characteristics visible under the microscope prove that it is made from accumulations of plant remains.

Still other sedimentary rocks, such as the salt deposits of the Bonneville flats west of Great Salt Lake, Utah, are residues from the evaporation of saline lakes—identical beds can be seen in process of forming in parts of the lake a few miles farther east.

Most sedimentary rocks form distinct layers or *strata*. This *stratification* (also called *bedding*) generally results from variations in the supply of sedimentary detritus during deposition, or from changes in the velocity of currents that are laying down the material, or from still other factors. Even a few minutes' observation of a sand bar in a small stream shows that such variations have taken place recently.

A smooth-walled pit dug a few inches into the sand bar usualy reveals stratification identical with that of many sandstones. Visits to an ocean beach during and after a storm reveal comparable changes in the coarseness of the beach material.

A pit dug by a child in the sand commonly shows layers of varying coarseness recording such variations in current and wave strength.

The strong waves and currents of a heavy storm may partially erode an earlier deposited layer of sand and mud on the sea floor and sweep a sheet of coarse gravel over it, as has been proved by samples dredged from the same spot before and after storms.

Even the accumulation of shells in an offshore reef is sometimes interrupted by a fall of ash and pumice from a nearby volcano or by mud swept far out to sea during unusually heavy floods in the rivers of the adjacent land. By such interruptions and accidents during deposition, distinct sedimentary strata are formed; and because such changes vary in intensity and frequency, some strata are thin, others many feet in thickness.

Laws of Sedimentary Sequence

Observations of strata now accumulating make possible the following rather obvious generalizations, which are useful in interpreting ancient sedimentary rocks:

In any pile of sedimentary strata that has not been disturbed by folding or overturning since accumulation, the youngest stratum is at the top and the oldest at the base.

In other words, the order of deposition is from the bottom upward. This is the *Law of Superposition*, first clearly stated in 1669 by Nicholas Steno.

Water-laid sediments are deposited in strata that are not far from horizontal, and parallel or nearly parallel to the surface on which they are accumulating.

Though the bottom surface of a stratum may conform roughly to the irregularities of the base, its top must be nearly horizontal. This is the *Law of Origirial Horizontality*, also stated by Steno.

Many applications of these generalizations will appear in subsequent that they are not insignificant truisms may be realized from the fact that in most mountain ranges the strata of sandstone, limestone, and other sedimentary rocks are no longer horizontal. Instead, they are steeply tilted or even overturned.

Because we know that these rocks were once sheets of sand, shells, and gravel deposited in nearly horizontal layers and then cemented together, their present distorted attitudes show that great forces must have been at work in the region.

Thus, a simple structure like the stratification of a sediment gives us the clue to read the record of great changes of energy in the earth—changes that have deformed and mashed once horizontal sheets of rock into fantastically complex patterns.

CLASSIFICATION OF SEDIMENTARY ROCKS

Sedimentary rocks are classified and named on the basis of their *texture*, that is, the *size and shape of their constituent particles*; and on their *composition*, that is, *the kinds of material that compose the particles and cements*.

Two general subdivisions may be recognized: *Clastic sedimentary rocks* are composed mainly of fragments of rocks and minerals which have been transported to the site of deposition and cemented there; *organic and chemical sedimentary rocks* are composed dominantly of the cemented shells of organisms, or of precipitates from aqueous solutions such as sea water.

Igneous Rocks

At active volcanoes, also, we can see rocks being made. Molten material rises to the surface and flows for a time in lava streams, but eventually cools and hardens into *igneous rock*.

Two subdivisions of the igneous rocks are recognized. The lavas and solid fragments erupted from volcanoes are called *volcanic rocks*. They are composed in large part of microscopic mineral crystals and glass. In a few places volcanic rocks can be seen to grade into rocks composed of much larger crystals.

These coarse-grained igneous rocks are called *plutonic rocks*, and are widespread at the earth's surface, but have not been observed in process of formation. There is good reason to believe that the plutonic rocks were not spewed out to the surface like the lavas, but solidified deep underground. Subsequently, the roof rock that covered many of these buried masses was eroded away, thus exposing the plutonic rock at the surface.

Volcanic Rocks

In January 1938, a white-hot stream of molten lava issued from a fissure near the base of the volcano Nyamlagira, in central Africa, and poured quietly downward into a forested plain below.

For two years and four months the lava continued to emerge until more than 500,000 cubic yards of molten rock had devastated an area of over twenty-five square miles. Finally the flow ceased, the fissure froze over, and the lava congealed into the black slaggy rock that we call basalt. Similar but smaller flows have been observed at many other volcanoes.

Some volcanoes erupt explosively, blowing vast quantities of volcanic "ash" and broken rock fragments into the air. Around volcanoes such as Mount Katmai in Alaska, Bandai-San in Japan, or Mount Pelée in the West Indies, large areas have more than once been buried under fifty feet or more of hot ash and rock. Such loose debris, erupted from Vesuvius in Italy, has been known to consolidate into firm, coherent rock within a generation.

That rocks were made by volcanic action was well known to the peoples of early civilizations because of the many active volcanoes in the Mediterranean countries and in Persia. Early writings contain a wealth of information on Vesuvius in Italy, Santorin in the Aegean Sea, Etna in Sicily, and Erebus in ancient Persia.

The eruption from Vesuvius that overwhelmed Herculaneum and Pompeii in 79 A.D. was but one of a series of volcanic disasters that impressed on early peoples both the awesome power of volcanic eruptions and the characteristics of the lavas and explosive products that come from them.

Nevertheless, the ancients did not recognize that volcanic rocks are widely distributed over the surface of the earth in areas far removed from volcanoes now active. It is one thing to stand on the brink of a fissure and watch liquid lava emerge, roll down a slope, and congeal into a mass of basalt, and another to recognize a basalt flow that was extruded millions of years ago, and has since been detached from its parent cone or fissure by erosion, or buried under a later accumulation of sedimentary rock.

From their manner of formation, it is common for volcanic and sedimentary rocks to be interstratified. In the Samoan Islands, flows of basalt have been seen to spread over reefs in which limestone was forming; today, deposits of coral and shells are collecting on the upper surface of the congealed lava.

The great flow at the base of Nyamlagira covers older volcanic material, but earlier flows in the same region spread out over a plain underlain by lake and river deposits and, in turn, were partially buried under new deposits from the lakes and rivers. Scarcely any thick pile of sedimentary rocks is completely free from volcanic interlayers.

It is not surprising that flows of lava and beds of volcanic ash interstratified with sedimentary rocks were considered to be hardened sediments by the early geologists, just as they often are by the uninitiated today. It was not until after much careful field study that reliable criteria were put forth distinguishing buried lava flows from sedimentary deposits. Indeed, fifty years of violent controversy took place before the volcanic origin of basalt was definitely established.

THE CONTROVERSY OVER THE ORIGIN OF BASALT

The interpretation of geological phenomena is often influenced by the philosophy of the worker. The history of geology is replete with examples of unsuccessful attempts to fit the features seen in the field into the preconceived notions of the observer. One of the classical examples of the conflict between theoretical and field interpretations was the controversy concerning the origin of basalt, a controversy that raged from about 1775 to 1822.

Some of the hills of Saxony near the famous mining academy of Freiberg (Bergakademie) are composed chiefly of sedimentary rocks, but interstratified with them are a few layers of a hard, dark-colored rock long ago named basalt. The basalt is more resistant to erosion than the sedimentary rocks with which it is associated and appears in picturesque colonnaded cliffs at or near the summits of many of the hills.

In 1775, the Stolpen, one of the basalt-capped hills, was visited by Abraham Gottlob Werner, professor of mining and mineralogy at Freiberg, a scientist who was destined to wield great influence on the development of geology ("geognosy," as he called it).

From his observations on this and later visits, Werner wrote, in 1787, that the hill showed "... not a trace of volcanic action, nor the smallest proof of volcanic origin. . . . After further more matured research and consideration, I hold that no basalt is volcanic, but that all these rocks, as well as the other Primitive and Floetz* rocks, are of aqueous origin." Following up his idea that all basalts were precipitated from the ocean, Werner developed a "system."

With all the precision that he applied to the classification and organization of the mineral specimens in the laboratory collections at Freiberg, he proceeded to divide the crust of the earth into a series of "Universal Formations." These, he taught, were all precipitated from a primeval ocean, and could be definitely recognized all over the world, each formation having the same character and occurring in the same order no matter in what country it might be found.

Werner's Personal charm was great, and he attracted large numbers of able students whom he fired with great zeal. They came to believe that the "Universal System" of the Master would unlock the geologic history of every country.

Some of them were rudely awakened; in fact, Wernerism received its first really serious setback from two of Werner's own students, D'Aubuisson and von Buch. Both of these young men became disillusioned with Werner's "system" after visiting the Auvergne region in central France. The Auvergne has had no eruptions within historic times but it contains almost perfectly preserved craters, lava flows, and other volcanic phenomena.

The Auvergne had been studied several years before by Nicholas Desmarest (1725-1815), a hard-working government official who in his spare time published several excellent papers on geology. Among these, his treatises on the Auvergne volcanoes have won for him the title of "Father of Volcanology."

How different was the approach of this thorough observer from that of the dogmatic Werner to essentially the same problem! In his first journey into the Auvergne in 1763, Desmarest found a cliff of basalt. Searching at the base of the cliff, he noticed that the soil beneath the flow had apparently been burned and hardened.

He also noticed that the basalt grades into and contains masses of *scoria*, a coarsely frothy basaltic rock filled with small, spherical holes. Scoria is common at the base, and even more abundant in the upper part, of basalt flows; it has been observed to form when rising bubbles of steam are caught by the congealing of the sticky lava around them.

When he first visited the Auvergne, Desmarest had never seen moving lava from an active volcano. By careful observation and by reasoning, bowever, he established two of the criteria now generally used in the recognition of ancient basalt flows—the baking of the ground beneath the flow and the presence of scoria formed by the congealing of bubble-filled lava. Desmarest, however, did not proceed forthwith to establish a "system."

His curiosity was aroused, but he drew no certain conclusions. He decided to examine and map the outlines of the entire flow. This work disclosed similar features at many other points along the base of the flow. He also traced the flow to its source at the base of a round, steep-sided bill which still retains the characteristic form of a volcanic cone.

Still not entirely satisfied, Desmarest decided to plot on a map the distribution of all the different kinds of rocks of the Auvergne. By carefully following the junctions between the lavas and other kinds of rocks, and by plotting these junctions or *contacts* on maps, he proved that the volcanic history of the Auvergne was very long.

Some eruptions had been followed by long quiet periods during which streams cut valleys in the flows and removed much of the ash from the cones. Then, these newly cut valleys were filled and obliterated by renewed volcanic activity. Eventually he traced out three main cycles in the volcanic history. His map, one of the first geologic maps produced, is a monument to his good judgment and ability to interpret field relations. Desmarest's maps will stand comparison with modern maps of similar volcanic districts.

Thus, nearly 200 years ago, the failure of Werner's speculations to withstand the test of Desmarest's rigorous field observations showed geologists that the ultimate worth of a theory can only be found in the field. Ironically, although most of Desmarest's reports were published before Werner tried to compress all the vagaries of volcanic action, sedimentation, and erosion into a "Universal System," they remained almost unnoticed for many years. While Werner's teachings and ideas were sweeping over Europe, Desmarest took no part in the controversy. When asked for his opinion on the origin of basalt, he would reply:

Eventually D'Aubuisson and von Buch, eager to establish the "Universal System" of their teacher in other countries, visited the Auvergne. How great was their disillusionment as they followed, step by step, the same evidence and lines of reasoning that had led Desmarest to his conclusion! Here in the Auvergne, D'Aubuisson and von Buch saw the burned and hardened ground on which the lava had flowed, and the scoriaceous tops and bottoms of the flows where rising gas bubbles had been trapped in the congealing lava.

All these features could have been seen at Stolpen had Werner taken his students there on field trips instead of merely lecturing about the "Universal System" and giving them specimens of basalt to study in the laboratory. Even more devastating for the Wernerian but nearby they also rested on granite, which supposedly occupied quite a different position in the "Universal System." The reports of D'Aubuisson and von Buch did much to overthrow the theory that basalt was a precipitate in the "Universal Ocean," although other students of the great teacher, still content to work in the laboratory instead of examining rocks in the field, continued to promulgate the Wernerian doctrines.

The controversy over the origin of basalt had happy results for the development of volcanic geology. Interest in ancient volcanoes spread wide and far and bore abundant fruit, particularly in the British Isles, an area uniquely rich in varied and spectacular volcanic phenomena, notwithstanding the fact that none of the British volcanoes has been active in historic time.

Plutonic Rocks

As the details of lava streams and deposits of volcanic ash became better understood, the British geologists turned their attention to the conduits through which the molten rock, or *magma*, as such natural silicate melts are called, had reached the surface. In the Midland Valley of Scotland and in parts of the Hebrides, erosion has swept away most of the lava streams and cones and has cut deeply into the foundation beneath.

Here, as in northwestern New Mexico and in central Oregon, erosion has displayed in clear and diagrammatic fashion the conduits that fed the volcanoes. Many of these feeders are simple, pipelike masses (*volcanic necks*) or filled fissures (*dikes*); but some connect with other subterranean igneous masses much more complex in form and origin.

Soon a wide variety of different kinds of igneous bodies formed by the solidification of magma underground had been mapped and described. These are called *intrusive igneous bodies* in contrast to the *extrusive bodies* formed by magma erupted to the surface. The rocks that have solidified within them are called *plutonic rocks*, in contrast to the *volcanic rocks* that have formed at the surface.

Observations of hundreds of different intrusive masses have shown that plutonic rocks are characterized by coarser mineral grains than comparable lava flows. This has led to the inference that the slower-cooling underground affords time for the mineral grains to grow larger. In large intrusions, some of which are many miles in diameter, the grains are easily visible, and the minerals can be readily identified without the microscope.

Laboratory investigations prove that heat is transmitted by solid rock less readily than it is carried off by rising air (convection currents) above a molten lava flow. Hence, the roof rock acts as a blanket which permits only slow cooling of a magma emplaced below the earth's surface.

Rounded gas bubbles are absent or at least very rare in plutonic rocks, apparently because of the considerable pressure of the roof which keeps the gas in solution in the magma. The gas collects into bubbles only when the pressure is released, as when magma is suddenly extruded from depth to a point upon or near the surface.

Although some intrusive bodies can be traced continuously into recognizable lava flows so that their origin from a common magma is obvious, most are separate bodies. These may never have had any direct outlet to the earth's surface.

That these masses were likewise formed by crystallization from a magma is proved by several lines of evidence: They commonly harden, or even completely recrystallize, the rocks along their contacts; tongues and stringers from them penetrate into cracks in the adjacent rock in the manner of a liquid; they commonly show chilled borders of fine textured rock similar to that of lava flows; and they display, with minor exceptions, the same mineral and chemical compositions as surface flows.

Most intrusive bodies cut across the bedding of the enclosing sedimentary rocks, hence they are called *discordant*, but a sill is a rock mass congealed from magma forced *concordantly* between sedimentary strata. The magma has spread out like the grease squirted between metal surfaces by a grease gun.

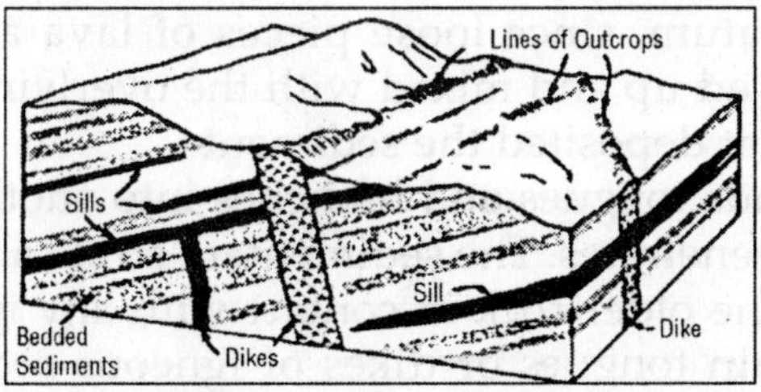

Fig. Sills are Corcordant, Dikes are Discordant Tabular Intrusion

A sill might thus occur low in the sequence and yet be much younger than strata hundreds of feet above it. Here, then, is a possible source of error in applying our generalization about the sequence of strata, for how can we tell an intrusive sill from a lava flow that has been buried tinder a later accumulation of sedimentary rocks?

CONTACTS OF ROCK MASSES

The key to this question, as to many others in geology, lies in the interpretation of the contacts between rock bodies. A mass of a single kind of rock, whether sandstone, basalt, or any other, does not extend indefinitely; somewhere it must end against another.

These common boundary surfaces of adjacent rock masses are called their *contacts*. There are two general kinds: *abrupt*, with a definite surface of junction; and *graduational*, in which there is no sharp boundary but an intermediate zone of greater or less thickness in which the transition from one rock mass to the other takes place.

In interpreting the relative age of two rock masses in contact, the following generalizations are useful. They apply not only to problems of sequence among igneous rocks, but to all kinds of rock bodies:

Of two rock masses in contact, that one is younger which contains within it fragments or inclusions of the other.

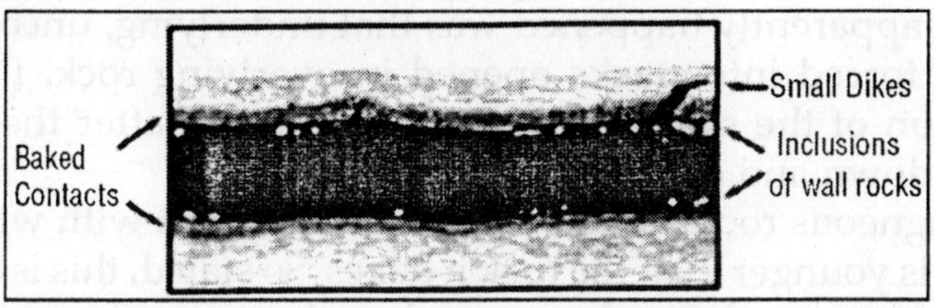

Fig. Criteria for Distinguishing Between a Sill and a Buried Lava Flow

Thus, a sill might be expected to enclose, at some places near its upper surface, fragments torn from the overlying stratum at the time of intrusion. On the other hand, fragments of a buried lava flow are likely to be found as inclusions in an immediately overlying sedimentary stratum, since loose pieces of lava and scoria would probably be picked up and mixed with the overlying detritus by the moving agent that deposited the sediment.

If a rock sends tongues and branches into another, it is younger than the rock it penetrates. The sedimentary strata above most sills, or for that matter the older rock in contact with any igneous intrusion, commonly contain tongues or dikes of igneous rock formed by the magma penetrating fissures or other openings in the older rock and solidifying there.

Although this rule applies mostly to igneous rocks, sediments may also, in places, penetrate adjacent rocks. Thus, cracks and openings in a lava flow may contain debris identical with that of an immediately overlying sedimentary rock. Presumably, some of the unconsolidated detritus filtered into the cracks during deposition of the sediment and cemented there into rock.

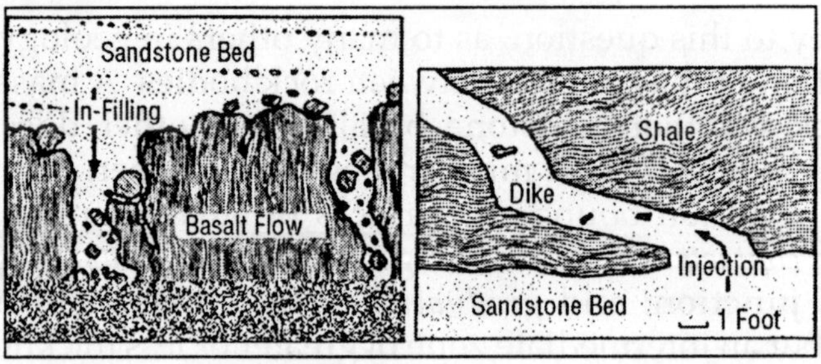

Fig. Sandstone Dikes

The generalization must be used with care, however, when applied to sedimentary rocks. For example, an underlying sandstone may send dikes upward into overlying shale. This is an exception to the rule as stated, though of course the rise of the sand into dikes actually occurred after the formation of the overlying shale.

What apparently happened was that underlying, unconsolidated sand was forced into cracks opened in overlying rock. Presumably cementation of the sandstone was delayed until after the shale had been laid down and consolidated.

If an igneous rock bakes or alters another rock with which it is in contact, it is younger than the rock it bakes. So stated, this is an obvious truism, but to recognize such contact alteration is not always easy. Some rock masses are exceptionally well cemented or impregnated with mineral matter along their contacts because of later percolating waters. Along other contacts, rocks have been discolored by water stains until they may look like chilled zones. Generally, careful observation or even microscopic study is needed to distinguish these spurious features from actual baking or chilling.

We have already seen how Desmarest noted the alteration of the rock beneath the Auvergne, lavas. A sill will alter not only the strata that form its floor, but also those of its roof. A moving lava flow has no roof, and lava congeals before any considerable sedimentary cover can be deposited upon it.

In general, the criterion of alteration a hot magma can be more readily applied to intrusive rocks than to lava flows. The interior of a lava flow commonly remains molten long after a thick rind of congealed lava has formed on its surface.

Many flows have been seen to move forward beneath this rind; during movement the still-liquid interior breaks up the solidified crust and rolls blocks of it under the front of the flow. Such blocks are commonly so cool that they fail to bake the material beneath—or at least baking is so slight that it is not readily recognized. But intrusive masses of igneous rock invariably alter their walls, at least to some degree.

THE INTRUSIVE NATURE OF GRANITE

As careful geologic mapping in many areas of volcanic and shallow plutonic rocks disclosed more and more about the contact relations of intrusive igneous rocks, a spirited discussion arose over the origin of granite, a common, coarse-grained rock composed chiefly of feldspar and quartz. The origin of granite is still a topic of lively discussion but many points have been clarified by careful study of granite contacts.

Geologic mapping in many different countries shows that granite and the similar rock, granodiorite, are among the most abundant rocks in the accessible part of the earth. They are found in great monotonous bodies hundreds of square miles in extent.

At many places, eroded granites are covered with sedimentary rocks that contain pebbles of the underlying granite and are therefore younger than the granite. Other bodies are covered by younger lavas and beds of volcanic ash. These relations persuaded some early geologists to regard granite as the earth's "original crust," formed when the earth solidified from a molten state, though, as we have seen, Werner considered it to be the first precipitate out of a universal ocean.

The idea that all granites are old, however, was not left unchallenged. In many localities, geologists began to note that granitic dikes, tongues, and complicated offshoots of considerable size penetrate the overlying rocks, indicating that the granite had injected the roof rocks as a molten mass. Only in scale of the injection features—in miles or thousands of feet rather than in feet or inches—were the contacts different from those of small intrusive masses. Furthermore, the invaded rocks at such contacts were altered, and, in places, so thoroughly recrystallized that it was difficult to tell whether they had originally been sedimentary or volcanic rocks. Such changes in the wall rocks suggested the effect of high temperatures.

More detailed field studies also showed that most of the great expanses of granite, far from being as monotonous as had been supposed, actually embrace several distinct intrusions of granite and related plutonic rocks, the younger ones penetrating the older. Also, along many contacts, fragments of the adjacent older rock are strewn through the margin of the granite in large numbers as though they had been pried off by the intruding granite magma..

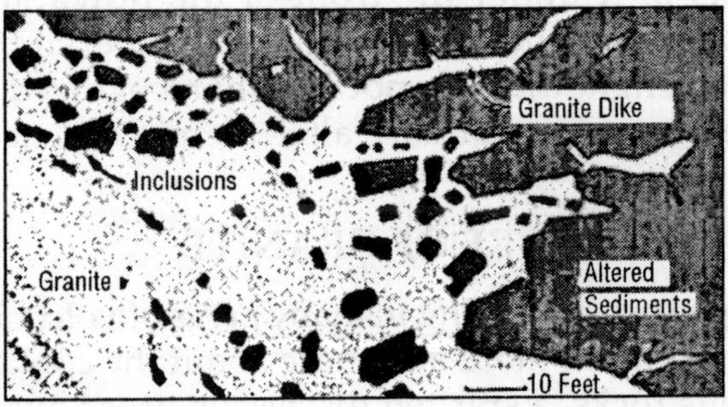

Fig. Intrusive Relations at a Granite Contact

Today, no remnants of the hypothetical "original crust" have survived the results of detailed mapping. At some points along their contacts, most granite masses can be seen to have invaded and altered rocks of sedimentary or volcanic origin, although elsewhere younger sedimentary and volcanic rocks may have been deposited upon the eroded granite surface.

Thus, many granites were shown to be igneous rocks, congealed at considerable depth. Diked contacts and other injection phenomena also indicated that some masses were not single intrusions but had been formed by successive invasions separated by intervals of inaction. This recalls the Auvergne, where Desmarest demonstrated three distinct effusive periods separated by long quiet interludes. Apparently the emplacement of large igneous bodies, whether at the surface or far beneath it, is a slow and somewhat intermittent process.

If we left the subject of the origin of granite here, however, we would be greatly oversimplifying a complex problem. Not all granite contacts are either clearly erosional or clearly intrusive. Many of them are gradational, the granite appearing to fade gradually into rocks of undoubted sedimentary or volcanic origin.

Also, certain rocks, definitely granites in composition and texture, contain faint, nebulous patterns that resemble stratification, outlines of pebbles, or other structures found in sedimentary or metamorphic rocks. The origin of these transitional contacts and of granites with ghostlike foreign structures is a fascinating problem that has not yet been completely solved.

Classification of Igneous Rocks

With the petrographic microscope, much information can be obtained about the minerals in an igneous rock, including their kinds, shapes, sizes, patterns of arrangement, order of crystallization, and the alterations brought about in them either by hot gases from the magma or by outside agencies such as air and atmospheric moisture. Great variation in so many different features yields infinite possibilities for classification, and so literally hundreds of kinds of igneous rocks have been discriminated and given separate names.

However, the great bulk (approximately 95 per cent) of the igneous rocks can be lumped into about 15 major groups. This simplified list suffices for our purpose; we shall not need names for the numerous varieties within each group, or for the rare groups not included. Such a simplified classification of the igneous rocks can be based on the minerals and textures visible to the eye, thus dispensing completely with the microscope. The classification outlined in Table IV-2 is widely used in field work where laboratory equipment is not immediately available.

Comparison of mineral composition and the bulk chemical composition of thousands of igneous rocks shows that the kinds and amounts of different minerals depend, in general, on the chemical composition of the original magma. Magmas rich in silica yield, on cooling, large amounts of feldspar and quartz; magma low in silica form rocks rich in the ferromagnesian minerals such as pyroxene, olivine, and amphibole.

Textures of Igneous Rocks

The size of the mineral grains in an igneous rock depends chiefly on the rate of cooling, although bulk chemical composition of the magma plays a part. It has been inferred from field observations, and confirmed by laboratory experiment, that a high content of water and other volatile substances promotes the growth of larger crystals.

Field observations clearly indicate that when magma congeals slowly in a large subterranean mass, the minerals commonly grow large enough to be readily identified with the eye.

If the magma forms a lava flow, however, rapid cooling in contact with the air prevents the growth of large crystals, and, instead, the rock congeals to mixtures of microscopic mineral grains and glass. If cooling is even more rapid the magma may congeal into a glass containing almost no crystals.

These differences in *degree of crystallinity* and in *size of crystals* determine the *texture* of the igneous rock. A rock is said to have a *granular texture* if its mineral grains have grown to a *size large enough to be seen and identified without the aid of lens or microscope.* In different rocks, the average size of the grains may vary from about 0.5 mm. to more than 1 cm. in diameter, but the common coarse granular rocks such as granite have grains averaging from about 3 mm. to 5 mm. in size. In rocks with *aphanitic texture,* the constituent minerals are less than 0.5 mm. in diameter; they are commonly mere specks, too small to identify without the microscope.

In many aphanitic rocks, the microscope shows a considerable residue of uncrystallized glass between the minute crystals, but others are completely crystalline.

Rocks with *glassy texture* are almost entirely volcanic glass, though even they may contain a few scattered crystals.

Pyroclastic texture is the texture of rocks formed of volcanic ash and rock fragments blown out of a volcano and cemented together in the manner of sedimentary rocks, or else welded together by heat from the chemical reactions that take place between the gases in the eruption cloud.

Many igneous rocks contain crystals of two widely different sizes and therefore appear spotted; these rocks are said to have a porphyritic texture and are often loosely called "porphyries". Because porphyritic texture is most common in small intrusive bodies or in lavas, it has been attributed to *a change in the rate of cooling while the magma was crystallizing*.

The process is hypothetically explained as follows: A large body of magma underground may cool to the temperature where one or more minerals begin to crystallize. Because cooling is slow, the crystals of these minerals grow to considerable size. If, when the magma is perhaps half crystallized, a fissure opens in the roof of the chamber, some of the magma with its suspended crystals may escape to form a lava flow at the surface.

The still-liquid portion of the magma quickly freezes and surrounds the large crystals, or *phenocrysts*, with a *groundmass* of crystals of aphanitic size. The phenocrysts were formed underground, the aphanitic groundmass at the surface. Such a lava has a *porphyritic aphanitic* texture.

Rocks with *porphyritic granular* texture—large crystals in a granular groundmass of finer grain—are common in some intrusive bodies. *Porphyritic glassy* texture is seen in some lava flows. Rarely, conditions other than a change in the rate of cooling may locally produce porphyritic rocks.

Metamorphic Rocks

We have seen that the sedimentary origin of many rocks can be confidently inferred from their stratification, water-worn pebbles, fossils, or other characteristic features. Other rocks display textures and mineral compositions that, by analogy with those of rock masses of known volcanic or plutonic origin, permit us to classify them just as confidently as igneous rocks.

There remains a third great group of rocks in which diagnostic features of igneous or sedimentary origin either are absent or have been so obscured and altered by the growth of new minerals and new textures that they are scarcely recognizable.

Thus, in places, we find rocks showing rounded pebbles and stratification much like that of normal conglomerates, but the pebbles are ellipsoidal instead of spherical; the matrix between them consists not of sand and clay as in ordinary conglomerate, but of lustrous foils of mica and quartz; and, cutting across two or more adjacent pebbles and the intervening matrix, are long delicate needles of tourmaline.

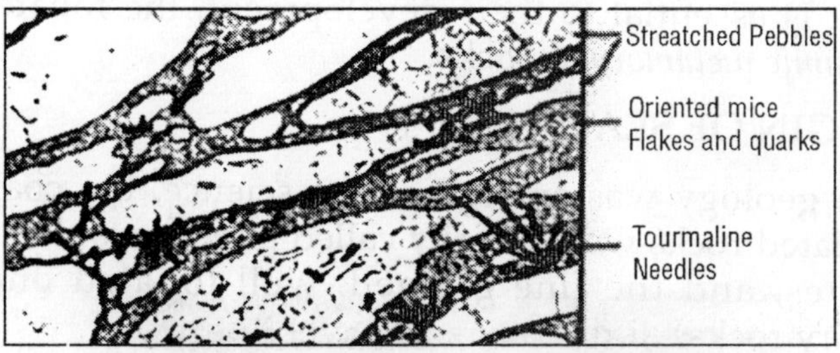

Fig. Metamorphosed Conglomerate with Tourmaline Needles Cutting the Other Constituents

It is inferred that the rock was once a conglomerate and that the tourmaline could not have been present when the conglomerate was deposited but must have grown in the solid rock long after deposition. Apparently the original round pebbles have been squeezed and stretched, and the sand and clay originally present in the conglomerate matrix have been altered to mica and quartz. Similarly, igneous features such as phenocrysts or flow banding may locally be recognized in their metamorphosed equivalents.

Still other rocks show no trace whatever of the textures of known sedimentary and igneous rocks. Yet, in places, they grade imperceptibly into rocks in which partially obliterated igneous or sedimentary structures still survive.

From such relations, the conclusion that this last great group of rocks represent altered or *metamorphosed* rocks is well established. We therefore recognize this third group as *metamorphic rocks*. There can, from the nature of their origin, be no sharp boundary between metamorphic rocks on the one hand and the igneous or sedimentary rocks from which they were derived on the other.

The survival of relic features such as pebbles or phenocrysts in metamorphic rocks must mean that the original rock did not melt in attaining the metamorphic condition. Instead, the alteration appears to have been accomplished mainly by recrystallization while the rock remained essentially solid.

Foliated Metamorphic Rocks

Most metamorphic rocks have a banded or layered structure called *foliation*. The layers may be coarse bands 1/24 inch or more thick, as in gneiss or layers thinner than a sheet of paper, as in slate. The rock usually splits readily along these planes.

The foliation appears to record widespread pervasive movement within the rock mass, during which most of the original minerals were broken, streaked out, and recrystallized into new minerals. Because movement is essential to their development, the foliated rocks are called *dynamic metamorphic* rocks.

THE ORIGIN OF SLATE

When geology was emerging as a science, the coarse-grained, faintly foliated rocks were usually called gneisses and classified with the granites, and the fine-grained, well-foliated ones with the sedimentary rocks.

For example, the metamorphic foliation in common slate, a rock much used for roofing, flagstones, and other building purposes, was erroneously considered to be stratification. Geologic study of European and American slate quarries disclosed, however, that most slates have two distinct layered structures.

One can be seen to cross and displace the other. The older of these is parallel with other interlayered rocks like limestone or quartzite, and with alternations in grain size, colour, and composition; it is the true sedimentary stratification. The rock, however, does not break or split along the stratification planes as in normal sedimentary rocks except when the younger structure, the true metamorphic foliation, happens to coincide with it.

When foliation and stratification do coincide, recognizable but distorted fossils are found locally on the

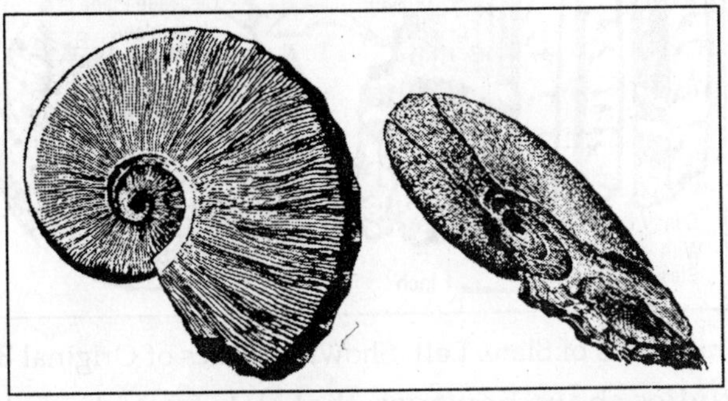

Fig. The Fossil Collected from Unaltered Limestone in Idaho

foliation surfaces. Where these fossils can be compared with like specimens from otherwise little-changed sedimentary rocks, they are seen to have been greatly thinned in the direction at right angles to the foliation and are stretched out parallel to the foliation. If the foliation cuts the stratification at a high angle, however, it is almost useless to look for fossils, because the rock will rarely split along the stratification planes where the fossils occur.

Furthermore, each foliation plane, where it crosses the stratification, commonly breaks and offsets the stratification planes by a small (usually microscopic) distance, showing that it is a structure younger than the bedding.

Thus, a fossil may be broken and displaced by so many microscopic shears as to be unrecognizable.

In most slate quarries the foliation is nearly parallel throughout, although the strata themselves may be thrown into folds and become highly distorted. The foliation cuts through the folds, intersecting the strata at all angles.

From the relics of stratification and the occasional battered fossils, it is clear that most slates were derived from fine-grained sedimentary rocks, such as shale. This is confirmed by chemical analyses which show no essential difference between shale and most slates, except that the slates contain less chemically combined water than the shales.

Field and laboratory studies show, however, that not every slate is derived from shale; some were formed from other fine-grained rocks such as volcanic tuff. Only by the use of the petrographic microscope and by X-rays can we unravel the true nature of the foliation. The minerals of slate are exceedingly fine grained—indeed, most are ultramicroscopic.

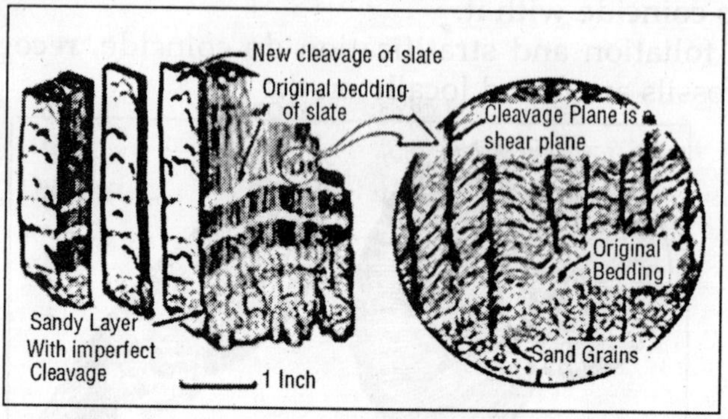

Fig. Fragments of Slate, Left, Showing Relics of Original Bedding

X-ray studies show, however, that slate is made up largely of tiny flakes of minerals akin to muscovite, the white mica. These flakes are nearly parallel. Since mica has only one perfect cleavage which is parallel to the platy form of each grain, the cleavage of all the grains in slate lies nearly parallel. Slate splits so readily into thin parallel sheets because of this cleavage parallelism.

Evidently slate is merely shale that has been sheared and heated sufficiently to cause the original clay minerals composing it to recrystallize into tiny micaceous flakes.

Movement along the innumerable surfaces of foliation, as shown by offsets of the bedding, has granulated the original minerals and has lined up the newly formed mica, which continued growing during and after movement.

In some areas, slate may be seen to pass by imperceptible gradation into more coarsely crystalline rocks such as mica schist. In these more highly metamorphic rocks, every trace of the original stratification and of fossils has been obliterated. Yet, in chemical composition they are identical to the slate into which they grade.

Although they may contain minerals different from those in slate, the new minerals in the transitional rocks can be seen to have grown from the micaceous minerals of the slate, gradually replacing and obliterating them just as the micas in slate replaced the minerals of the shale.

From such mineral relations and transitions in rock type, we conclude that even these rocks, showing no trace of original sedimentary structures, were, in fact, derived from the sedimentary rock shale. They are merely more thoroughly metamorphosed.

Recrystallization at Igneous

One of the simplest kinds of metamorphism is the recrystallization of rocks by an igneous intrusion. In many places where, for example, pure quartz sandstone containing marine shells has been invaded by a mass of gabbro or some other plutonic rock, shells cannot be found in the rock close to the contact. Instead, the sandstone near the intrusion contains sheaves of a light colored mineral called wollastonite.

The wollastonite formed by the reaction has no resemblance to either the quartz grains or the shells from which it was derived. Most fossiliferous sandstones, however, contain many substances besides quartz—usually a little clay, some limonite or ferromagnesian minerals, and various other impurities. These generally join in the reaction, yielding, not a simple mineral like wollastonite, but a more complicated compound, usually a member of an isomorphous series such as amphibole, garnet, or pyroxene.

Where wollastonite actually does appear, however, we can make certain additional inferences about the conditions of metamorphism because the wollastonite reaction has been carefully investigated over a wide range of temperatures and pressures in the laboratory.

The results of this investigation are plotted. From this graph, we infer that if a rock containing wollastonite was formed at depths equivalent to a pressure of 3,000 atmospheres, the temperature attained by the metamorphosed rock was at least 850° C. This illustrates what is meant when certain minerals are called *"geologic thermometers."*

Locally, *contact metamorphism*, as the kind of metamorphism caused by the heat of igneous masses is called, is greatly complicated by the addition of various substances from the igneous mass. That such additions have taken place is demonstrated by the finding of rocks of different mineral and chemical composition along the contacts—for example, calcium-iron silicates taking the place of calcite in limestone near a granite contact.

These different minerals are attributed to hot gases and aqueous solutions escaping from the magma and carrying large amounts of iron, silica, sulfur, and other substances into the wall rock. Here they combined with the minerals already present to produce new minerals.

Recrystallization of the Stassfurt Salts

Deep wells show that the temperature in the outer crust of the earth increases, on the average, about 1° Centigrade for every 100 feet of depth. Therefore, if the rocks are buried to great depths, some of their constituent minerals may become unstable at the higher temperatures.

An interesting example of such *geothermal metamorphism* is afforded by the well-known salt deposits at Stassfurt, Germany. These salt deposits were studied in 1900-1905 by the famous Dutch chemist van't Hoff. Because of the associated sedimentary rocks, he reasoned that the salts had been formed from a shallow body of sea water, cut off from the ocean by a bar or other obstruction, and slowly evaporated in a desert climate.

Van't Hoff was puzzled by one feature of the salts, however. Although their bulk chemical composition is consistent with their being the residue left from the evaporation of sea water, the actual minerals of the deposits are quite different from those obtained by evaporating sea water to dryness.

Patiently and thoroughly, van't Hoff worked out the physical chemistry of the salts by many laboratory experiments, and, ultimately, produced many of the minerals found in the Stassfurt salts by holding sea water at temperatures well above normal surface temperatures but below boiling during the period of evaporation.

Van't Hoff then announced that the average annual temperature of the earth's crust at the time when the salts were laid down must have been much greater than at present—perhaps close to the temperature of boiling water!

Geologists, though accepting van't Hoff's experimental data, could not agree with the conclusion. Fossil fish and mollusks are found in strata associated with the Stassfurt salts. It is inconceivable that the marine animals of that ancient time had lived in hot seas. From the relations shown on a detailed geologic map of the Stassfurt area, it is obvious that, after the Stassfurt salts had been laid down in an arm of an ancient sea, they were buried under several thousand feet of younger sedimentary rocks.

The rise in temperature because of this deep burial could have caused the already precipitated salts to recombine into new minerals stable in the hotter environment. The textures of the salt, as seen in thin sections under the petrographic microscope, give abundant evidence of post depositional recrystallization of the solid salt beds. Thus, van't Hoff's conclusion as to the temperature of mineral formation was correct; his inferences as to the climate were absurd because he neglected the field evidence.

Salt beds are particularly susceptible to metamorphic changes; they recrystallize completely at temperatures that are still too low to cause notable changes in the silicate minerals of which most rocks are composed. In sandstones associated with the Stassfurt salts fossil shells are completely unaltered.

Therefore, the rise in temperature was clearly not enough to cause calcite to combine with quartz and form wollastonite. It would have required much deeper burial to metamorphose the sandstones and shales at Stassfurt; only the sensitive salt beds were affected.

METAMORPHIC PROCESSES

Even these brief descriptions—and scores of additional examples could be given—show that many minerals are chemically stable only within a limited range of pressure and temperature. If brought into a part of the earth's crust where the temperature is higher, or where crushing and shearing take place, or where hot fluids attack them, many minerals break down and form other minerals stable in the new environment.

Metamorphic rocks result from such transformations. In some, the alteration affects only a few minerals of the original sedimentary or igneous rock, and the metamorphic rock retains in the surviving unchanged minerals or textures clear evidence of its origin; but more commonly, complete transformation has occurred.

From relations revealed by geologic maps, we know that nearly all metamorphism takes place deep within the crust of the earth—far below the depths we can reach in mines and wells. Therefore, in general, we must infer the reactions that take place from the resultant products, although we can also get valuable clues from experiments made possible by laboratory equipment designed to imitate conditions deep within the earth. Unfortunately for our purpose, knowledge of the chemistry of solids, and particularly of solids under shearing stress, is still meager as compared with the vast amount of information chemists have obtained on reactions in solutions.

SIGNIFICANCE OF FOLIATED METAMORPHIC ROCKS

Where large areas of metamorphic rocks-particularly the foliated rocks called crystalline schists—are found at the surface, deep erosion has taken place. Such a generalization is based on observed relations of metamorphic and nonmetamorphic rocks in many parts of the world. We infer from these relations that the textures and minerals characteristic of foliated metamorphic rocks were formed at great depths, under conditions of high pressure and temperature.

Thus, we also interpret other rocks of like mineral composition and texture as having been under like conditions—even where the local structural relations do not permit direct proof of deep burial. We reason that a thick covering must have been eroded from any area where crystalline schists are now seen at the surface.

CLASSIFICATION OF METAMORPHIC ROCKS

Since any rock may be recrystallized by metamorphism and there are several different ways in which metamorphism may take place, the number of different kinds of metamorphic rocks is large. It is possible, for example, to find at least five different metamorphic rocks all with the chemical composition of basalt—and indeed, all derived from basalt—yet each differing from the others in texture, mineral composition, and general appearance.

A greater variety of minerals appears in metamorphic rocks than in either sedimentary or igneous rocks. Therefore, metamorphism is best studied with the petrographic microscope, aided by the principles of physical chemistry and, of course, primarily by many field observations.

Metamorphic rocks, like sedimentary and igneous rocks, are classified on the basis of texture and composition. The principal textures to learn for determinations made without the aid of the microscope are: Gneissose Texture. Coarsely foliated; individual folia 1 mm. or more, even up to several centimeters thick.

The separate folia are commonly lens like or wavy, and generally differ in composition; feldspars, for example, commonly alternate with dark minerals. Mineral grains are coarse, easily identified. Schistose Texture. Finely foliated, forming thin parallel bands or flat lenses. Individual mineral grains are distinctly visible. The minerals are mainly platy or rodlike—chiefly mica, chlorite, and amphibole. Equidimensional minerals like feldspar and pyroxene are not abundant.

Slaty Texture. Very fine foliation, producing almost rigidly parallel planes of splitting due to the parallelism of microscopic and ultramicroscopic grains of platy minerals such as mica. Granoblastic Texture. Unfoliated, or only faintly foliated. Mineral grains are large enough to be visible. Corresponds roughly to the granular texture of igneous rocks. Hornfelsic Texture. Unfoliated. Mineral grains commonly microscopic to ultramicroscopic, though a few may be visible. Corresponds to the aphanitic texture of lavas, from which it can be distinguished only with difficulty.

SECOND-RATE PLANET

Among the nine planets that speed endlessly around the sun is this terrestrial sphere. It is neither the largest nor the most massive, and astronomers are apt to remind us not only that it is a second-rate planet but that our sun itself is only a third-rater among the other stars. But because the earth is our abode, to us it is of supreme importance.

Fig. Terrestial Sphere

One of the first things we learn about the earth is that it is round. Some of us may even remember that a simple proof is that large objects gradually disappear below the horizon as one moves away from them. This is at best only a qualitative argument, but there are several precise proofs. As an example of the scientists' method of attacking such problems, let us follow through the steps involved in determining both the shape and size of the earth.

A little study of a circle will show that the property of circularity involves certain relations between the length of an arc of the circle and the angle made by the radii limiting this arc. Note in the accompanying figure that the radii OA and OB establish the arc AB. This arc is so related to the angle a that if we choose another are equal in length to AB, such as CD, then the angle b must be of the same size as angle a.

In the same manner, an arc twice the length of AB will be subtended by an angle twice the size of a. We may now argue that if every section through the earth is a circle, then equal distances on its circumference will be subtended by equal angles made by the radii from the centre of the earth to the ends of the arcs. In order to measure these internal angles, we must make use of astronomy. Let us, therefore, imagine two observatories about 2,100 mi. apart along a north south line. In each observatory we shall suspend a plumb line, which is nothing more than a string with a weight on the end.

This line will point to the centre of the earth if the earth is round. Now at the first observatory a telescope is pointed at a star directly overhead. At the same instant, the telescope in the second observatory is pointed at the same star; but because the second observatory is some

2,100 mi. from the first, the star obviously will no longer be directly overhead. Thus there will be an angle between the plumb line and the direction of the telescope. This angle is measured, and it is found to be about 30°.

We now make a diagram like the adjacent one. At A is the first observatory, with the plumb line and the telescope pointed directly overhead at the star. At B is the second observatory, with its telescope also pointed at the same star, but with an angle between the telescope and the plumb line.

Note, however, that the two telescopes themselves are parallel, because the distance to the star is so great that the rays of light reaching the earth from it are practically parallel at any point on the earth.

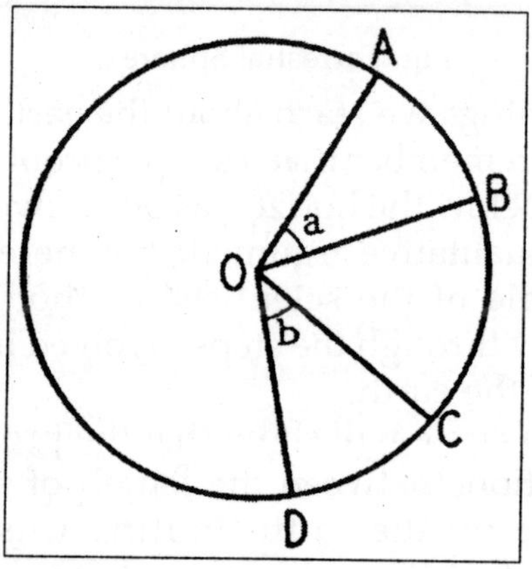

Fig. Angle a is the Same as Angle b, Provided Arc AB = arc CD

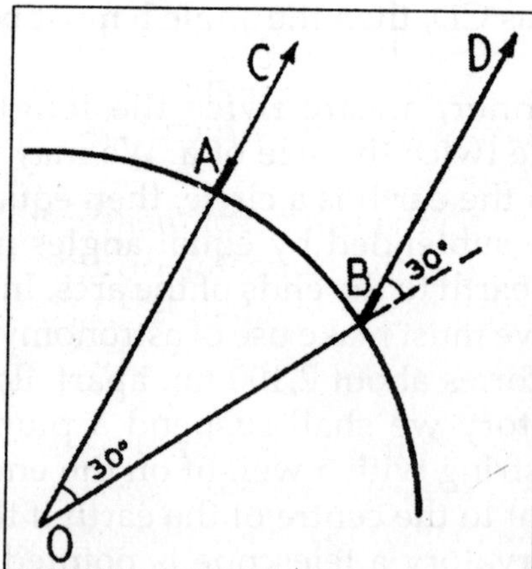

Fig. Two Observatories 2,100 mi. Apart Simultaneously Observe the Same Star

Remembering that the angle between the telescope and the plumb line of the second observatory is 30°, we next project the plumb lines into the interior of the earth, where they meet. The angle made by their junction is also 30°, because the lines AC and BD are parallel, and the line OB cutting these parallel lines makes an equal angle with each. Thus, it follows that an angle of 30° at the earth's centre subtends a distance of about 2,100 mi. along the surface. If we next divide through by 30, we see that an angle of 1° must subtend a distance of some 70 mi.

Should the same experiment be repeated elsewhere on the earth, and the net result prove to be the same, then we might safely conclude that the earth is spherical. Such measurements have actually been made at many places over the surface of the earth, and everywhere the general relations hold, so that there is no doubt that the earth is spherical.

Precisely speaking, of course, the earth is a spheroid, and flattened slightly at the poles. This flattening is so slight that between the polar and equatorial diameters there is a difference of only 27 mi., which is less than one-half of 1 per cent. It is, in fact, so insignificant that it amounts to only about ½ in. on a 10-ft. globe, and consequently is not shown even on the largest of such representations of the earth.

We have by these experiments found not only that the earth is spherical but also that every degree of the interior angle subtends about 70 mi. on the surface. Since there are 360° in a complete circle, we need only multiply our 70 mi. by 360 to obtain the *circumference*, or distance around the earth.

When this is done, the result is approximately 25,000 mi. Once we have the value of the earth's circumference, we can easily find its *diameter*, or the distance through the earth. We simply divide the circumference by 3.1416 (the value of the old familiar *pi* of school days) and the answer is about 8,000 mi., which is the approximate diameter of the earth.

As long as we are on the subject of measurements, let us pursue it a bit farther. If we know the diameter of the earth, we can find its *volume*, because the volume of any ball is 3.1416/6 times the diameter cubed. If we put in our known values, we find that the volume of the earth is nearly 260,000,000,000 cu. mi.

Similarly, since the surface of a sphere is equal to *pi* times the diameter squared, we may compute the *surface area* of the earth. Such a computation shows us that there are about 197,000,000 sq. mi. on its broad surface. But large as this area may seem, it is only about one twelve-thousandth the area of the sun's great surface. In other words, the earth's surface area is as much smaller than the sun's as tiny

Switzerland is smaller than the earth's entire extent of land and sea. Another important item concerns the *mean density* of the earth, or its mass per unit of volume. To find this, it is necessary first to determine the mass of the entire earth and then to divide this by its volume. The mass of the earth can be determined by several methods, but perhaps the simplest is that known as the Cavendisk experiment.* By this procedure the attraction of two large lead balls for two small silver balls is measured, and then compared to the attraction between the earth and one of the smaller balls.

From the exceedingly delicate measurements made in this experiment it is found that the mass of the earth is 6×10^{21} tons, which is the number 6 followed by 21 ciphers. To obtain the mean density of the earth, then, we divide the mass by its volume; whereupon the density is found to be approximately 5.5. In other words, a cubic foot of average earth-substance weighs about 5.5 times as much as a cubic foot of water.

But right here we stumble onto an interesting thing. The average density of the familiar rocks of the earth's surface is only about 2.7. Thus the density of the earth's interior must be greater than 5.5, in order to counterbalance the lightness of the surface materials.

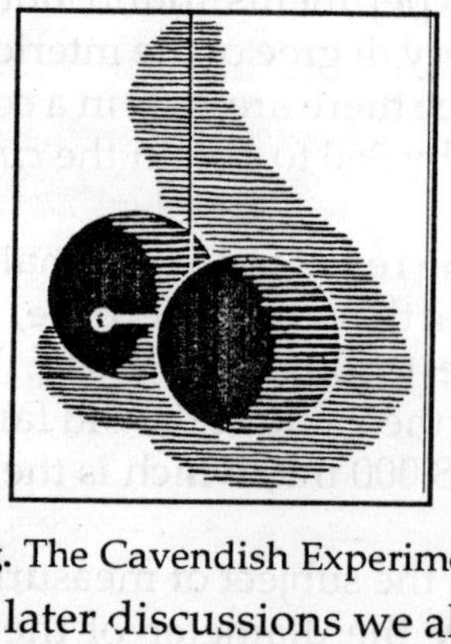

Fig. The Cavendish Experiment

For the purposes of later discussions we also ought to have before us a fairly complete picture of the earth's motions in space.

Most proofs of the *earth's rotation* are difficult to grasp without some knowledge of physics. The one usually cited depends on the fact that a heavy pendulum tends to swing in a fixed direction in space as long as no outside forces disturb it. Evidently, then, if such a pendulum is set into motion on a rotating earth, the earth should rotate beneath the pendulum without affecting the direction of swing of the pendulum.

Such, in fact, is the result of the pendulum experiment. When first set into motion, it is directed so that the bob swings in a north-south plane. As time goes on, it is found that the pendulum continually departs from this direction, and that in the Northern Hemisphere the plane of swing tends to rotate in a northeast-southwest direction.

It must be borne in mind, however, that an observer on the earth has the same motion as the earth, and therefore he will feel himself at rest and will notice that the direction of swing of the pendulum appears to change.

This departure would increase indefinitely were it not for friction stopping the swing of the pendulum. This experiment was first performed by Foucault in Paris, in 1851, and in his honour the device is called the Foucault pendulum. The theory of the Foucault pendulum is entirely beyond the scope of this book, but it is interesting to note that observation and theory agree so well that there is no doubt of the actual rotation of the earth upon its axis.

The motion of the earth about the sun, or the *revolution* of the earth, also can be demonstrated in several ways. One of the simplest is by the periodic shift in the position of the nearby stars with respect to the more remote stars. In the adjoining figure the earth's orbit around the sun, S, is shown diagrammatically.

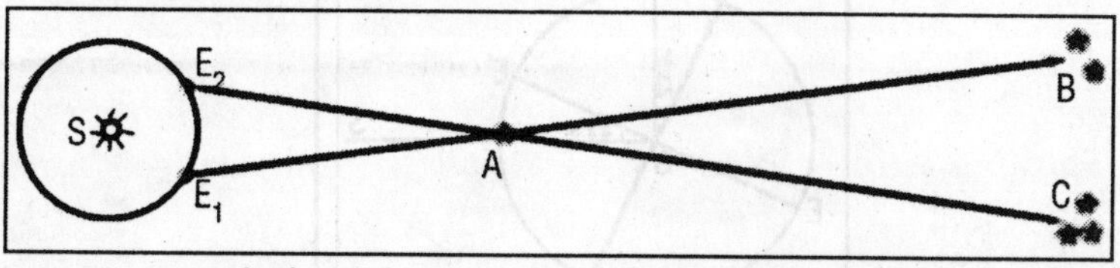

The apparent shift of the nearby star proves the earth's revolution. The sizes and distances are obviously not true to scale self appears in two successive positions, E_1 and E_2, several months apart in its journey around the sun. Star A is one of the nearer stars; and in the distance are two groups of stars, B and C, representing more remote constellations.

Notice that when the earth is at E_1, the star A, as viewed from the earth, appears to be in group B, but that later, when the earth is at E_2, star A appears to have moved to a position among group C. Thus star A periodically appears to shift back and forth between groups B and C. Such shifts actually are noted among the nearer stars, and the phenomenon is cited as the parallax proof of the earth's revolution.

The *inclination of the earth's axis* to the plane of its orbit is another feature of this second-rate planet into which we should briefly inquire. This inclination is demonstrated by the fact that at any point on the earth's surface the altitude of the sun above the horizon at noon on June 21 and December 21 differs by 47°.

This apparent change in the altitude of the sun measures twice the angle of inclination of the earth's axis, which thus turns out to be 23½°. In the adjacent figure the line OS, from the centre of the earth to the centre of the sun, lies in the plane of the orbit. Now at O let us erect a perpendicular OA, and then draw the earth's axis, OP at an angle of 23½° to OA. This angle, POA, measures the inclination of the earth's axis.

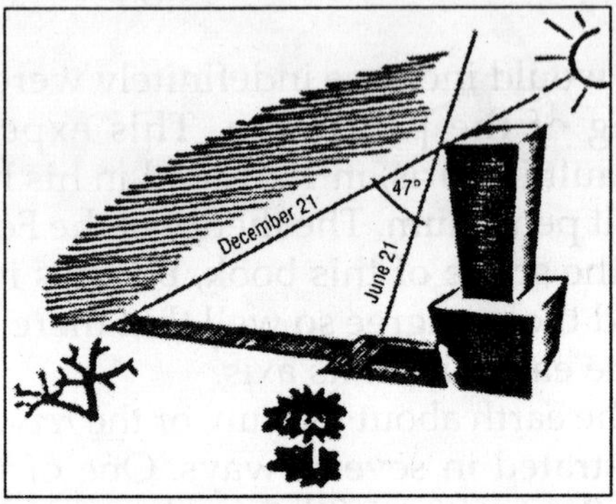

Fig. The Angle of the Sun's Rays Differs by 47° in June and December

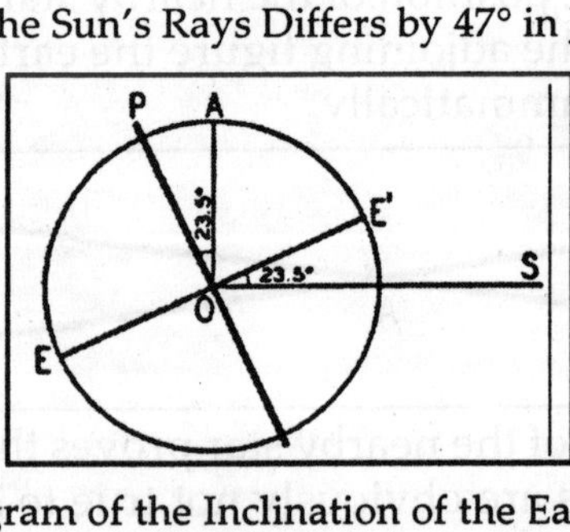

Fig. Diagram of the Inclination of the Earth's Axis

Inasmuch as the Equator of the earth is at right angles to its axis, it follows that the equator is inclined 28½° to the plane of the earth's orbit. This is line EOE_2.

It is the inclination of the earth's axis that causes the seasonal changes on the earth, as the figure below shows. The earth, like any rotating body, tends to keep its axis always oriented in the same direction in space, regardless of the earth's position in its orbit.

Notice, then, that when the earth is at the right in the diagram, the region receiving perpendicular rays from the sun lies 23½° *north* of the Equator; whereas when the earth is at the left, the perpendicular rays strike the surface 23½° *south* of the Equator.

In the course of a year, then, the sun seems to swing through an angle of twice 23½°, or 47°, in its apparent motion above and below the Equator.

When the earth is at the right, it is summer in the Northern Hemisphere, because at that time the northern half of the earth is receiving the more nearly perpendicular rays from the sun. In like manner, summer in the Southern Hemisphere and winter north of the Equator occur when the earth is on the left.

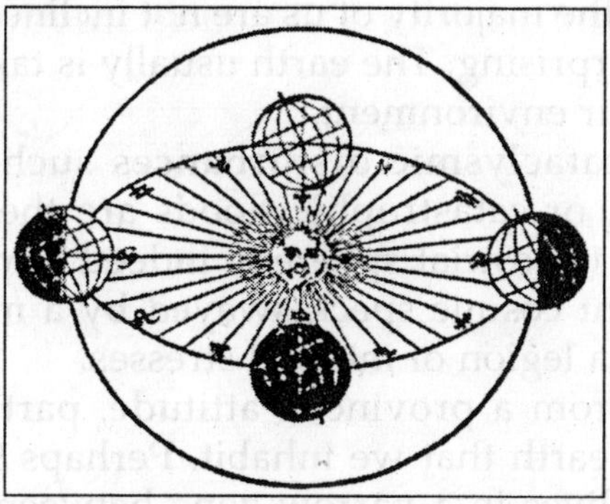

Fig. The Earth in its Journey Around the Sun

THE several points which have been discussed furnish us with the modicum of astronomical data that we need as a background for topics to be presented later. There are, of course, many other interesting features about the astronomical relationships of the earth, and the curious reader is urged to consult an elementary book on astronomy for details.

Item	Value
Shape	Very nearly spherical
Average circumference	24,800 mi.
Average diameter	7,910 mi.
Volume	2,599 × 108 cu. mi.
Surface area	197 × 106 sq. mi.
Mass (weight)	6,1 × 1021 tons
Average density	5.5 times that of water
Period of rotation	24 hr.
Period of revolution	3651/4 days
Inclination of axis	231/2°

By this time the reader should have in his mind's eye a fairly clear picture of the earth as a sphere rotating daily on its axis, like a fowl on a spit, as it annually revolves about the warming fire of that unfailing solar hearth, the sun.

Second-rate though it may be, this terrestrial sphere of ours has taken part in the stately procession of day and night, summer and winter, since time immemorial; and judging from its astronomical relations, it bids fair to continue into a future immeasurably long.

SOLID, LIQUID, AND GAS

The scientist thinks of the earth as an insignificant spheroid reeling through space, but the majority of us are not inclined to think about it at all. This is not surprising. The earth usually is taken for granted as the stable part of our environment.

Only during cataclysmic disturbances such as earthquakes, volcanic eruptions, or catastrophic floods are the unthinking ever reminded that this terrestrial sphere is indeed a mobile earth; is, in fact, an unimportant cosmic speck swayed by a myriad of external forces, wracked by a legion of internal stresses.

We all suffer from a provincial attitude, partly because of the limited part of the earth that we inhabit. Perhaps 90 per cent of the earth's population lives in a narrow zone between sea level and. ... thinks of the earth as an insignificant spheroid reeling through space .... about 1,000 ft. above the sea. This is an extremely thin zone, when contrasted with the 4,000 mi. of the earth's radius. Thus it may be well to pause here a while and attempt to acquire a more cosmopolitan view of the world as a whole.

It is difficult for us to visualize the earth in its entirety. Even extensive travel on its surface does not help. It is only through the study of maps or globes that its larger features may be understood at a glance. Let us, therefore, take an ordinary geographic globe—say 12 in. in diameter—and look at it from a distance of about 2½ ft. We will then be seeing the earth, on a proper scale, as if from the porthole of some super stratospheric rocket ship distant about 20,000 mi. from this terrestrial sphere.

Let's not be disappointed at the smoothness of the globe; the world's highest mountains themselves are not very impressive from such a distance. In fact, on the scale involved in a 12-in. globe, the highest points above sea level, some 6 mi., would be about 1/100 of an inch high. Thus, Everest itself is scarcely a bleb in the thin film of varnish that covers our little man-made geographic globe.

As we watch the earth for 25 hr., it turns completely around, and we readily observe that most of the surface is covered with water. The continents occupy only about one-fourth of the earth's surface, or only 54 million of the globe's 197 million square miles. One of the features of the earth that we cannot see from our vantage point is its envelope of air. Shifting clouds, however, would suggest the presence of an atmosphere and blot out some of the surface features. Hence these armchair observations have some advantages over those made from a rocket ship.

But coming down to earth, and landing safely, we may recognize three states of matter comprising the globe as a whole. The main solid body of the earth is composed of rocks and soil, and is called the lithosphere. Partly surrounding this solid body is the hydrosphere, composed largely of the oceans, which keep some three-fourths of the solid earth from contact with the atmosphere, the layer of gas that surrounds this entire terrestrial sphere.

The lithosphere itself is known only to a very shallow depth. No wells or mines penetrate much more than 2 mi. below the surface, nor were any of the rocks now exposed on the earth ever more than a relatively few miles below the surface. In fact, on the scale of our geographic globe we have direct knowledge of scarcely more than the layer of varnish on the surface.

Although we are inclined to think of the oceans as being very deep, and the atmosphere around us as extending for considerable distances upward, they are mere films wrapped around the solid body of the earth. The average depth of the oceans is about 2 mi.; but this is an insignificant part of 4,000 mi., the earth's radius.

In similar fashion, even if the true atmosphere extends out for as much as 150 mi. from the earth's surface, nevertheless this value is only about 4 per cent of the earth's radius. Thus we may safely speak of the air and water *films* about the earth, because on a world-scale they are little more.

But insignificant as these features might be to an astral being, they are quite impressive to us. Viewed from our limited zone on the earth, we justly admire mountain-climbers or stratosphere flyers who reach dizzy heights, or those intrepid explorers of the oceans who descend a half mile or more into the deeps. We shall accordingly return to our customary restricted environment and look at some of the relief features of the earth in more detail.

The major features of the lithosphere are the continents and ocean basins. If we include the shallowly submerged continental shelves with the continents, then we may say that about one-third of the earth's surface is occupied by the continental elevations and the remainder by the ocean basins. An idea of the differences in altitude between the ocean bottoms and the land surfaces may be had from the fact that the greatest oceanic depths are over 30,000 ft., or about 6 mi., and the highest mountains are nearly that high above sea level.

Thus there is a maximum relief of over 60,000 ft., or roughly 12 mi., on the lithosphere. Viewed provincially, these are large figures;

but in the light of our new cosmopolitan knowledge, we readily understand that the highest mountain and the deepest abyss are truly insignificant irregularities on the relatively smooth surface of this terrestrial sphere.

To help us visualize the distribution of irregularities on the earth's surface, let us consider the adjoining chart, called a *hypsograph*. The vertical scale expresses elevations above or below sea level, and the horizontal line represents the total area of the earth. The curve itself traces the distribution of relief on the globe. Notice how relatively little area is occupied by either the high mountains or the greatest oceanic depths.

By far the greatest portion of the continents is within a few thousand feet of sea level; and most of the wide expanse of ocean bottom is between 10,000 and 20,000 ft. below sea level. Some of the relief features of the earth are illustrated in Plate 1, to which the reader is referred.

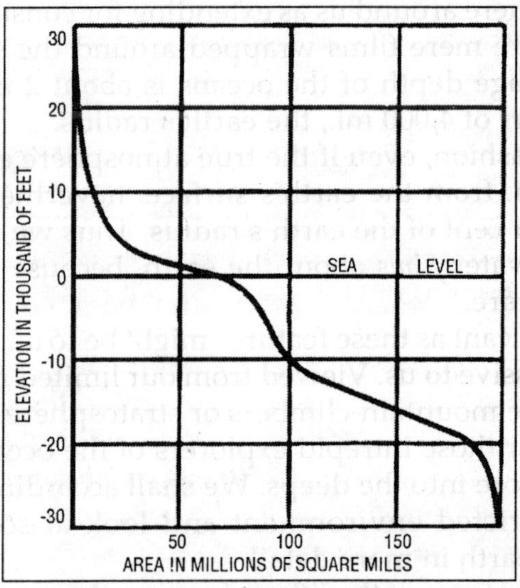

Fig. Hysograph Showing the Distribution of Relief on the Earth's Surface

Land features on the continents themselves may be divided into mountains, plateaus, and plains. Mountains are high lands of small summit area, prominent above their surroundings. Plateaus are high lands with considerable summit areas, rising abruptly above their surroundings on at least one side. Plains are low lands with little relief. In addition there are the smaller relief features, such as hills and valleys, familiar to all.

By and large, the major physical features of the earth are pretty well known to everyone. Not so, however, the chemical nature of the several Paris of the earth. Most people have learned at one time or another that the atmosphere consists mainly of oxygen and nitrogen. It also contains, in addition to the rare gases, small amounts of carbon dioxide and water vapour, which play very important rôles in geological processes.

Every high-school boy knows that water, and hence the hydrosphere, is a chemical compound of hydrogen and oxygen. When we consider the chemical composition of the lithosphere, however, we find that common knowledge is much more meager. And yet there are many important and interesting features of the solid part of the earth into which we should inquire.

As a first approach to the subject, let us consider two rather broad types of environment that may be recognized in the lithosphere. At the surface the pressure and temperature are comparatively low. Beneath the surface of the earth, however, and particularly at depths of 10 or more miles, the temperatures and pressures are relatively high.

The pressure effect can be understood easily when we remember that every cubic foot of rock must exert a pressure on the rock beneath it. A cubic foot of common surface rock weighs about 150 lb., and 10 mi. down the 52,800 cu. ft. of rock exert a pressure at the bottom of the column equivalent to thousands of pounds to the square inch, in contrast to the mere 14.7 lb. exerted by the atmosphere at the earth's surface. Likewise the temperature increases with increasing depth, and at a depth of 10 mi. it certainly is several hundred degrees higher than at the surface.

The point to this discussion of environments is that the kinds of chemical compounds that tend to occur in the lithosphere depend in large measure on the environment. We usually speak of the surface environment on the one hand, and deep-seated environments on the other, meaning by the latter an environment beneath the surface at a depth great enough to involve a considerable increase in temperature and pressure.

Now it happens that many rocks are formed deep beneath the surface of the earth, and as long as they are in their original environment of high temperatures and pressures they remain unchanged. Thus we speak of such rocks as being "in balance," or in *equilibrium*, with their environment. Broadly speaking, the deep seated environments tend toward the production of complicated chemical compounds, whereas the surface environment is more suitable to simpler compounds.

The rocks which are formed deep beneath the surface may be

brought nearer to the surface through the operation of geological changes, and thus they pass from a deep-seated environment to the surface environment. Obviously, of course, our direct observation is confined pretty well to surface conditions, but in deep canyons we may find types of rock from far below.

Likewise there are large areas, where rocks have been brought to the surface by geological changes and may still be found largely unaltered by surface conditions.

It is from such rocks that we reconstruct and interpret the conditions of the deep-seated environments, those great laboratories in which the rocks of the earth are largely made. Furthermore, direct experiments in our chemical and physical laboratories, made at high temperatures and pressures, supplement and confirm this evidence from the rocks.

A great many samples of all kinds of rock have been collected and chemically analyzed, so that we may confidently say something about the average composition of the outer part of the lithosphere. Tables have been prepared of the ultimate composition of the lithosphere in terms of the percentages of the elements present.

Element	*Percentage*
Oxygen	46.46
Silicon	27.61
Aluminum	8.07
Iron	5.06
Calcium	3.64
Sodium	2.75
Potassium	2.58
Magnesium	2.07
	98.24

When we look at such a table, we find to our surprise that out of the ninety-two elements, only eight are present to the extent of 1 per cent or more by weight.

Let us look here at the part of such a table that includes the eight most abundant elements. An outstanding feature of this table is that oxygen is by far the most abundant element of all. We do not mean to say that the oxygen exists in its free state. Obviously, it is chemically bound up in the rocks. Similarly with the seven other elements: not one of them occurs in its elemental state in the rocks; all of them are combined in one way or another into chemical compounds.

At this point some of us shall probably have to take a deep breath and hang on for a moment, because it may be easy to fall by the wayside. But if you are at all familiar with these elements, you know that, with the exception of oxygen and silicon, the other six elements are true metals; silicon has some metallic attributes, but in its properties it lies between the true metals and the non-metals.

All seven of the solid elements, however, do form definite oxides, and the suggestion is strong that the fundamental chemical unit with which we have to deal is the oxide. At any rate, if we talk about oxides we shall find that the discussion is much simplified.

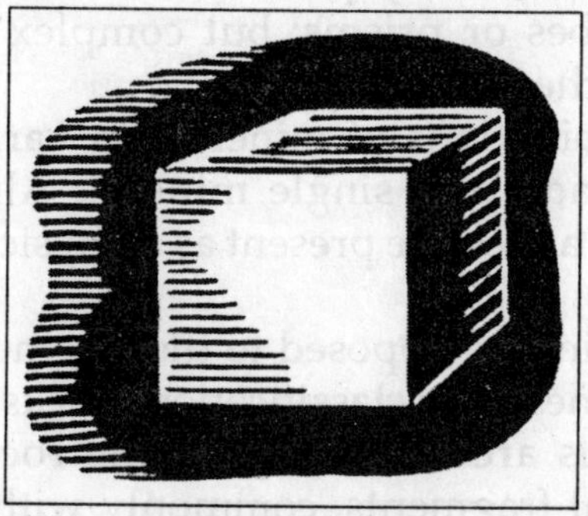

For example, metallic oxides tend to form *bases*, whereas the oxides of the non-metals tend to form *acids*. Silicon oxide shares this non-metallic attribute, at least in the presence of metallic oxides, and the result is that silicon oxide has a strong tendency to combine chemically with metallic oxides to form a silicate. For example, if magnesium oxide and silicon oxide combine chemically, a compound called *magnesium silicate* results. In chemical shorthand the reaction is expressed thus:

$MgO + SiO_2 \rightarrow MgSiO_3$

The resulting compound is one of the kinds usually formed under deep-seated conditions. Commonly, more than one of the metallic oxides combines with the silicon oxide to form double or triple silicates, like *calcium aluminum silicate,* $CaAl_2(SiO_3)_4$, which may be written $CaO.Al_2O_3.(SiO_2)_4$. It is thus a combination of calcium oxide, aluminum oxide, and silicon Oxide. Now these very silicates are the naturally occurring minerals that we find among the rocks of the lithosphere.

A mineral is a naturally occurring chemical compound; and although we may not be aware of it, we already are familiar with several minerals. They commonly occur as crystals, that is, in definite geometric forms bounded by plane surfaces. The simpler types of crystals may be cubes or prisms; but complex forms occur, as the accompanying sketches show.

Rocks are combinations of minerals in various proportions, or occasionally are made of a single mineral. Almost invariably the minerals making up a rock are present as a physical mixture of crystals or crystal fragments.

It is the crystalline, as opposed to the fragmental, nature of rocks that largely determines their classification. Rocks composed mainly of interlocking crystals are called crystalline rocks; whereas others, composed, of crystal fragments, commonly with broken or rounded edges, are called fragmental rocks.

The fragmental nature of this second type suggests that they were formed from the breaking-down or wearing-away of the crystalline rocks. This is so in fact but the story is no simple one. Indeed, we shall spend the next dozen the geological changes to which the crystalline rocks are subjected to form the fragmental rocks.

We shall begin our story with the minerals contained in the crystalline rocks, because these rocks are associated with the deep-seated environment. Geologists recognize literally thousands of minerals among natural earth-products.

But do not despair. Kind Nature has decreed that if we are familiar with some half-dozen of them, we shall be able to talk intelligently about most rocks.

Let's go back for a moment to the eight elements we mentioned. If we make a list of oxides that are formed from these elements, we have the following:

Silicon oxide, SiO_2	Calcium oxide, CaO
Iron oxide; FeO	Magnesium oxide, MgO
Iron oxide, Fe_2O_3	Potassium oxide, K_2O
Aluminum oxide, Al_2O_3	Sodium oxide, Na_2O

Iron forms two oxides (ferrous and ferric), while the other elements form one each. Some of the silicon oxide combines with other oxides to form silicates. Out of the welter of such silicates that are known, only two groups are just now important from our point of view. They are the ferromagnesian minerals, and the feldspars. In addition to these silicate minerals, silicon oxide itself commonly occurs as a mineral, in which case it is called quartz.

Look for a moment more closely at the mineral groups we have just listed. As the name of the first suggests, it is composed largely of iron and magnesium oxides. Actually, a number of individual mineral species are included in the ferromagnesians, but they have so much in common that we may well group them under this one term.

In general the ferromagnesians are chemical combinations of magnesium oxide and iron oxide with silicon oxide, and in addition they may contain calcium oxide and aluminum oxide. If you are interested in actual chemical formulas, we may list one of them which is fairly typical of the group: $Ca(Mg, Fe)(SiO_3)_2$. This is a calcium-bearing iron-magnesium silicate, commonly called *pyroxene*. The ferromagnesians are usually dark green to black in colour, owing to their high iron content.

The feldspars, in contrast to the ferromagnesians, are generally light in colour. They range from white through cream to pink and pale green. The feldspars are combinations of the oxides of aluminum, calcium, sodium, and potassium with silicon oxide, and they form rather complex silicates. The chemically minded may be interested in knowing that one common feldspar has the formula $(K,Na)AlSi_3O_8$, and another is $CaAl_2Si_2O_8$. The first of these is a sodium-potassium aluminosilicate called *orthoclase,* and the second is a calcium aluminosilicate called *anorthite*. Both are naturally occurring minerals.

We shall be quite chagrined if our readers feel it necessary to memorize any of these complicated chemical formulas; they are included merely as a shorthand method of indicating which of the eight most abundant elements they contain. Whenever we have occasion to mention these minerals in a chemical sense we shall point out the elements again, and thus refresh your mind.

However, if you are one of those hardy folk who insist on remembering formulas, then remember the one that we shall introduce next. The ancient Greeks believed that the mineral quartz was water frozen so firmly that it would never thaw again. This belief may have arisen from the finding of crystal-clear quartz in the Alps, among the perpetual snows that swathe those majestic peaks.

We are more prosaic nowadays, and we recognize quartz as silicon oxide itself, the chemical formula for which is SiO_2. Quartz, as a matter of fact, is one of the most abundant minerals on earth. The sand of beaches and streams is practically all quartz, broken and ground into fragments by the forces of nature. The abundance of quartz suggests that in nature there is an excess of this oxide above what is needed to form silicates with the metallic oxides.

The minerals we have been discussing occur most commonly in a particular group of the crystalline rocks, formed far beneath the surface of the earth, at high temperatures, and usually from liquid rock. Everyone is familiar with the phenomenon of volcanism, at least to the extent of knowing that there are vents in the earth's surface from which molten rocks pour.

This lava comes from beneath the surface, obviously; but much molten rock never sees the light of day. It is such lava, buried beneath the surface, protected from rapid cooling by a layer of other rocks above, that cools and crystallizes into rocks which contain the primary minerals we have cited. Plate 2 illustrates these primary minerals and shows some of their more common characteristics.

Before taking up in detail the changes that these crystalline rocks undergo to form the fragmental rocks, we must concern ourselves with one other important aspect of the physical set-up of the earth: the earth's sources of energy, and the passage of that energy through the economy of this terrestrial sphere.

ENERGY AUDIT

Business corporations periodically issue financial reports to indicate their solvency, and to suggest, perhaps, the desirability of investing in their stocks and bonds. A complete report includes not only a balance sheet but also a statement of profit and loss which summarizes the income and expenditures of the firm. A study of the sources of income and the channels of expenditure generally yields a clear picture of the nature of the business and the efficiency with which it is operated.

Now this terrestrial sphere has sources of income and channels of expenditure of energy. Suppose we constitute ourselves a planetary commission to audit the earth's income and outgo of that extremely important commodity. To do so, we must first consider the sources of energy in and contrast with them the come outgoing channels. We should also consider the manner in which the incoming supply of energy is distributed over the earth; and finally we should set up a statement of profit and loss, and note whether the balance is to be drawn in black ink or red.

The first and most obvious source of energy income is the sun. With the aid of suitable instruments and appropriate calculations, astronomers have been able to determine that the earth receives 1.94 cal. per min. per sq. cm. This is called the *solar constant* and, except for small variations, it appears to remain fairly uniform over long periods of time.

The stars, too, send us radiant energy; but when we attempt to evaluate it in the same terms as that of the sun, it is found to be an extremely small amount—so small, indeed, that it is to be compared with the augmented income that would accrue to the Standard Oil Company for selling an additional gallon of gasoline a year. For the purposes of our audit, then, we may quite safely ignore the energy income from the stars, as far as its recognizable effects on the earth are concerned.

In addition to its one main external energy source, the earth has a supply of internal energy. When one descends into a mine, or lowers a thermometer into a deep well, he finds that the temperature rises at a fairly uniform rate toward the earth's interior. On the average, this increase of temperature is about 1°C. for every 100 ft. of descent. The name geothermal gradient is applied to this downward increase of temperature.

Everybody knows that if we place a hot object in contact with a cold one, heat will flow from the hotter body to the colder. This flow of heat expresses itself to our senses as temperature decreasing away from the hot object; and, indeed, we can easily tell which way the heat is flowing by moving in the direction of decreasing temperature. If we apply this simple test to the earth, we find that the temperature decreases toward the surface, so that the heat itself must be flowing outward from its interior.

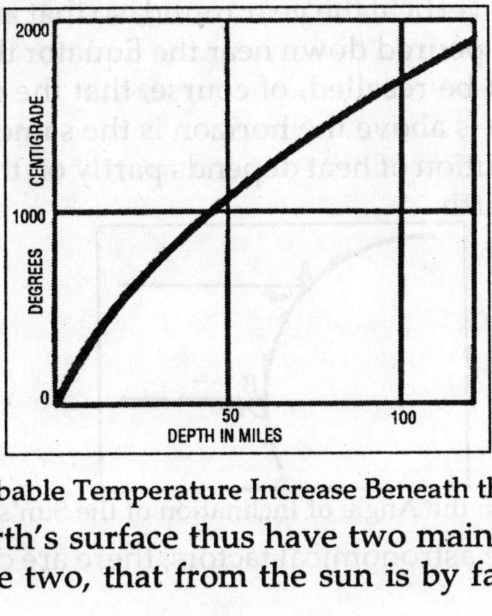

Fig. Graph of Probable Temperature Increase Beneath the Earth's Surface.

We on the earth's surface thus have two main sources of energy income; and of the two, that from the sun is by far the greater. But, although the heat radiated outward from the earth's interior is small in comparison with the energy received from the sun, it is nevertheless very much larger than the energy received from the stars; and it is of great importance in geological processes such as volcanism and the deposition of many of the ores.

Let us now consider how the energy income of the earth is distributed. If we confine our attention first to the external source of energy, we may see that the astronomical relations of the earth have much to do with it.

The earth's spheroidal shape, the inclination of its axis, and its movements on its axis and in its orbit around the sun cause this external energy to be distributed over the earth's surface in varying quantities: a variation from place to place and from time to time.

Consider for a moment some point on the earth's surface, such as Chicago, which is situated about 42° north latitude. During the day the sun warms the surface and raises the temperature.

At night some of the heat is radiated off into space and the temperature drops. Thus does the rotation of the earth affect the energy relations.

In summer the sun is more nearly overhead at noon, and its rays pass almost vertically through the atmosphere. In winter the noon sun is lower down toward the southern horizon, and hence its rays must follow a longer route through the atmosphere, with a consequent greater absorption. The result is that in summer more energy is poured directly on Chicago than in winter.

The spherical shape of the earth is brought into the picture when we contrast Chicago with some point nearer the Equator. There the rays are more nearly vertical the year round, so that annually a greater quantity of energy is poured down near the Equator than farther north at Chicago. It should be recalled, of course, that the number of hours per year that the sun is above the horizon is the same for all latitudes; but the actual distribution of heat depends partly on the angle at which the rays strike the earth.

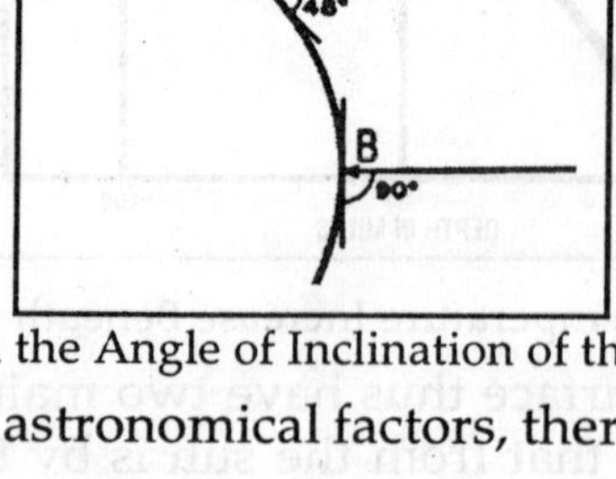

Fig. Effect of Latitude on the Angle of Inclination of the Sun's Rays on the Earth

In addition to the astronomical factors, there are certain terrestrial features that affect the distribution of the sun's energy over the earth. The amount of absorption by the atmosphere increases with its content of water vapour, carbon dioxide, dust and several other minor constituents.

Clouds reduce the amount of sun's rays directly striking the earth's surface, and differences in the nature of land and water determine largely the increase of temperature that follows the absorption of a given amount of sunlight on the surface.

The principal factor that controls temperature rise (assuming a given amount of absorbed heat and a given mass) is the *specific heat* of the substance.* Water requires about five times as much heat to raise its temperature a given amount than does an equal weight of rock. Thus with a given amount of absorbed sunlight, land areas will rise in temperature rapidly, while water areas will not. Of the heat which strikes the surface, some is directly reflected, of course, and so does not contribute to the temperature rise.

There are certain aspects of the sun's energy that we should consider. Much of the solar energy which penetrates the atmosphere is converted to heat, and it is almost entirely as heat that it is utilized on the earth. A number of important processes depend on this supply of heat and its distribution, so that it will be pertinent to pause here for a moment to examine them.

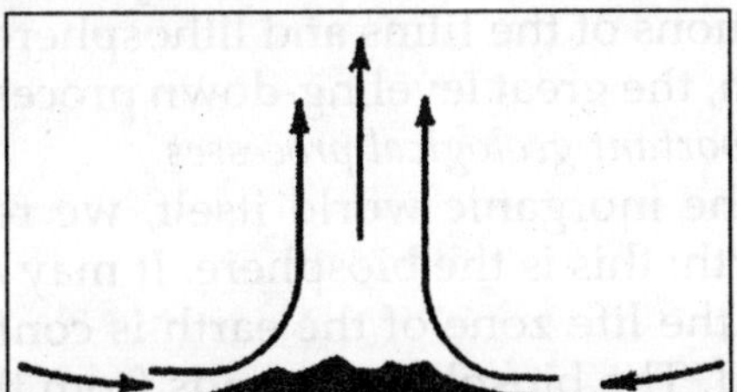

Fig. Cool Air Moves in and Displaces the Warmed Air

The distribution of the sun's energy over the earth, and the wide fluctuations in its amount at any given time or place, result in a differential heating in the earth's air and water films. Since heated air expands and becomes lighter, colder and heavier air may flow into the area of heating to displace the warmed air.

Thus are set up atmospheric circulations. In addition to the circulation of the winds, there is a constant passage from liquid water to gaseous water vapour, owing to evaporation from the surfaces of seas and lakes. This water vapour is swept along with the winds and eventually condenses back to water, falling as rain or snow. The rain feeds and swells the streams, which ultimately carry the water back to the ocean basins. Thus there are continual circulations within the films, and as the winds and waters flow over the earth they modify its surface. Everyone has seen examples of these air and water movements in wind storms that sweep clouds of dust with them, or in cloudbursts that swell small streams into torrents and load them with mud and débris.

All of these circulations require energy. As winds blow, they dissipate some of their energy in friction; and from other causes as well, they gradually lose their energy and die down.

Similarly, evaporation requires energy, which is stored in the vapour as latent heat, only to be released again when the vapour condenses. We begin to see here some of the devious paths followed by the energy that reaches the earth, but it is too early yet to summarize the process.

The circulations of the earth films cause a number of important phenomena known as geological changes. Air and water, or ice, acting on the solid part of the earth, modify its surface, by carrying away the transportable rock débris, in such manner that the irregularities of the surface, like hills and mountains, are slowly worn away and leveled down.

The material carried off is eventually deposited in the seas, those great receiving-basins of the earth. Thus the two films serve as active geological agents, and the surface of the lithosphere is the object acted upon. These interactions of the films and lithosphere are summarized in the term gradation, the great leveling-down process of geology. *This is the first of three important geological processes.*

In addition to the inorganic world itself, we recognize a fourth sphere about the earth; this is the biosphere. It may also be likened to a thin film, because the life zone of the earth is confined within very narrow limits indeed. The biosphere extends from the ocean bottoms to a point several miles above the earth's surface, and that is all.

Yet this limited zone is extremely important in the distribution and use of the earth's supply of energy. Plants absorb sunlight directly and convert it into potential chemical energy, which is stored in such substances as cellulose and starch. Animals may then eat the plants and use this energy for their own life-processes.

Part of the plant material may, however, be buried beneath the surface of the earth as coal, and there the potential energy remains locked up until released by the magic wand of fire. In similar fashion organic matter may be buried and later give rise to valuable pools of petroleum, with their high stores of potential chemical energy. Eventually, however, most of the energy stored up by plants and animals is released again through the slow processes of decay, and thus contributes to the outgoing energy which no longer figures largely in the economy of the earth.

We are now almost ready to prepare our statement of energy profit and loss, and we shall do so as soon as we have looked a bit more closely at the earth's interior sources of energy. Just as we do not know certainly what the ultimate sources of the sun's energy may be, so are we uncertain about the cause of the earth's interior heat. We may, however, reason intelligently about the matter.

There is a strong likelihood that a considerable part of the energy is released by *radioactive decay*. In fact, chemical analyses of common rocks indicate that their radioactive content is sufficient to account for a good portion, if not all, of the heat escaping from the earth's interior. It seems likely, however, that other sources of heat are also present. For example, we know that within the earth there are tremendous pressures due to the weight of layers and layers of rock from the surface downward.

The possible compaction of the earth's interior that may result from this could well generate heat in the struggle between the resistance of the earth body to shrinkage and the indomitable forces of compression. Whatever the sources the heat is nevertheless conducted always to regions of lower temperature; and in the earth this means toward the surface.

We have ample evidence, however, that all the energy Within the earth is not merely dissipated as heat. As a matter of fact, it is by no means certain that the heat being conducted outward from the interior of the earth represents even the major part of the earth's internal energy. The great compaction within the earth may there set up stresses and strains; and when the limits of adaptability are reached, the earth yields by deformation.

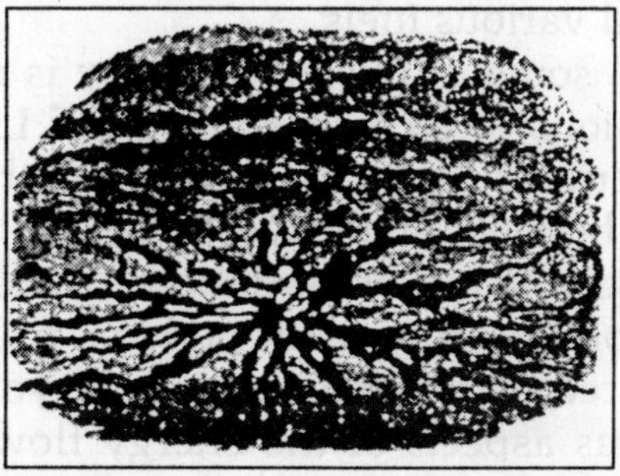

Fig. Earthquakes are an Evidence of the Earth's the Movements are Very Very Slow, Internal Energy

We have ample evidences of earth deformation; *earthquakes* are an example. Such deformations or movements of the solid part of the earth, either rapid or slow, are called diastrophism. Usually adjustments, and when they occur, the local area is shaken by earthquakes. Serious as they are from a human viewpoint, earthquakes are thus merely shivers and spasms within the earth body.

Diastrophic events may express themselves in many ways, and later on we shall devote a series For the present we may point out that in all its aspects diastrophism involves energy in some form or other, and thus belongs in our ledgers. So important is diastrophism in earth economy that it is called the *second great process of geology*.

The energy of the earth's interior tends to be radiated outward into space, but of course it must first be conducted to the surface. During this conduction, large bodies of rock may become heated above their melting-points, despite the great pressures, and so liquefy. Movements of this liquid rock. or lava, within and upon the earth, give rise to volcanism, the *third great process of geology*.

If the liquid rock finds access to the surface, we have the spectacular phenomena of volcanic eruptions. If it solidifies again underground, the results are no less far reaching, but of course not so spectacular, for, although the outpourings of lava on the earth cause havoc and destruction, the more quiet intrusions of lava among the rocks below the surface commonly give rise to valuable and extensive ore deposits.

We now have all the data we need to complete our audit. We saw that the incoming energy arose from two sources—the sun and the earth's interior. In the transformation of the sun's radiation through the energy scheme of the earth, the short incoming light waves are largely converted to heat energy, and as such are used in the activities of wind, water, and animal life, or are stored as potential chemical energy in plants and various fuels.

Simultaneously, some of this heat energy is dissipated outward into space as radiations of long wave-length.* Likewise the energy from the earth's interior is partly used in diastrophic adjustments and in volcanic activity; but eventually it, too, escapes as heat radiation. It thus seems that there is today essentially a balance between the incoming and outgoing energy, with perhaps a slight excess for the surplus account, stored up as potential energy in one form or another.

All these various aspects of the energy flow are shown in the appended diagram, which represents our condensed statement of profit and loss. It may not resemble an accounting record as much as an electrical circuit, but that is a minor matter.

The essential thing is that it furnishes us with a picture of the ebb and flow of energy upon and through the earth, and demonstrates that, although the percentage of profit may not be very high, nevertheless this terrestrial sphere is still a going concern, and certainly is in a position to continue actively in business for many years to come.

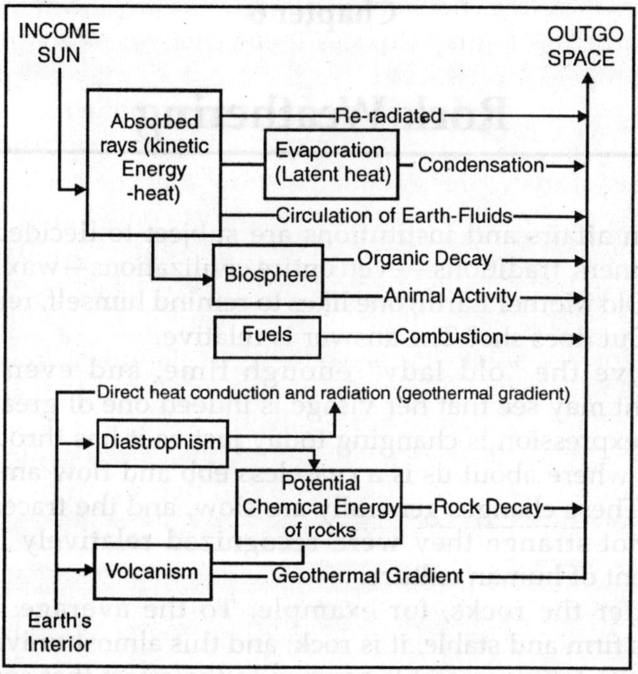

Chapter 6

Rock Weathering

Human affairs and institutions are subject to decided changes. Styles, manners, traditions—even entire civilizations—wax and wane. But Good Old Mother Earth, one likes to remind himself, remains ever the same. But does she? The answer is relative.

Just give the "old lady" enough time, and even the most unobservant may see that her visage is indeed one of great mobility. Her facial expression is changing today just as it has throughout the past. Everywhere about us is a ceaseless ebb and flow among earth-materials. These changes generally are slow, and the traces slight, so that it is not strange they were recognized relatively late in the development of human culture.

Consider the rocks, for example. To the average layman, if anything is firm and stable, it is rock; and this almost universal belief has given rise to innumerable sayings to the effect that something is "hard as rock." And yet there is abundant evidence that even rocks are subject to changes: that under surface environments they tend to break down into débris. In fact, this gradual decay of rocks is common knowledge, when one really stops to think about it.

Everyone knows that the writing on old tombstones is gradually obliterated by the tooth of time, and the blocks of most old stone buildings show signs of weakness and decay. Pitted surfaces develop, dark stains appear, or the blocks crack and crumble away. Few of us, however, have considered the implications of such slow changes extending over long periods of time. If marked changes occur within a few score of years, how much more apparent will they be in thousands of years.

Geologists have accumulated abundant evidence that under surface conditions rocks are relatively easily altered. The process of alteration is called rock-weathering. The nature of this weathering process depends largely on the kind of rock and on the specific conditions of its environment. In some places the surface of the rock becomes dull and stained, and pitted and crumbly.

In others the surface of the rock remains reasonably fresh, but thin shells break from its surface and joints and cracks appear in the body of the rock. The first kinds of change are chemical; the second are mainly physical. Both are included under the general term "weathering."

The net effect of most weathering changes is that hard and firm rock ultimately becomes a mass of small fragments and grains, which in most cases accumulate around the base or on the surface of the parent rock. This loose débris on the earth's surface is called mantle rock.

Because it consists mainly of small particles, it is subject to the mechanical forces of wind and running water. These agents may sweep the particles away from their parent location and deposit them elsewhere to form a second generation of rocks. In this sense weathering is the first stage in the ever continuing processes that wear down the land surfaces.

During the discussion of shallow and deep-seated environments it was pointed out that quartz and the silicate minerals (ferromagnesians and feldspar) were in equilibrium with their environment as long as that environment remained one deeply buried beneath the earth's surface.

But if the rocks containing these minerals are later exposed to the atmosphere by some geological change, the minerals which were formed under high temperatures and pressures are no longer permanent. In fact, the silicate minerals are relatively unstable and tend to break down to simpler compounds when exposed to the chemically active gases of the atmosphere, which are *oxygen, carbon dioxide,* and *water vapour*.

A complete analysis of rock-weathering deserves an entire volume, and so we must confine ourselves to certain fundamental notions. The principal points that we shall emphasize here are that:

- Weathering is an approach toward a state of equilibrium under surface conditions.
- This approach toward a stable state usually involves the breaking-down of the rock into finer débris.

Rock is a poor conductor of heat; and when the sun's rays pour down on exposed rocks, the surface layer is heated and expanded more than the interior. This expanding shell sets up strains within the rock and weakens it. At night the reverse process takes place, and the outermost layer contracts more than the inside.

If these expansions and contractions continue for a long enough time, the rock may finally yield by shelling off its outer layer. This loosened shell then fails from the parent block and lies at its base. Thus is exposed a fresh surface, ready for a repetition of the process. In detail this phenomenon is probably much more complicated than we have indicated, and involves some chemical changes as well as a yielding to physical forces. Time is undoubtedly an important factor.

Fig. Rocks Crack and Crumble under the Attack of the Weather.

Another aspect of temperature change is the disrupting effect of ice formed in rock crevices. All rocks, however dense in appearance, are penetrated by crevices and cracks, and into them rain water percolates. If the temperature drops sufficiently, the water freezes, and on freezing expands to about eleven-tenths its former volume. If there is no direction in which the expansion may take place freely, the ice exerts a pressure against the rock that may be great enough to disrupt it.

These are examples of physical weathering. Such physical changes take place most rapidly when high altitude is combined with steep slopes, as on mountain sides. Here the extremes of temperature are more pronounced, and the broken fragments rapidly fall away from the parent rock.

The blocks and fragments accumulate at the base of the mountains and build up talus slope, which reach up the sides of the mountains and may ultimately completely bury them. The reader is referred to Plate 3 for several photographs of dominantly physical weathering phenomena.

Simultaneously with the mechanical or physical weathering of rocks, chemical agents attack them and produce various decomposition products. When rocks are weakened and broken by physical forces, the resulting fragments have a total surface area much greater than that of the original rock exposure.

The speed of chemical reaction depends in part on the intimacy of contact between the reacting bodies, the greater surface area becomes an important factor in decomposition. The active gases of the atmosphere are dissolved in rain water which percolates among the fragments of broken rock. Thus these agents are able more effectively to perform their work of chemical decomposition.

Fig. Talus Slope at Base of the Twin Sisters

An important distinction between *physical* and *chemical* weathering must be made clear. Although physical disintegration changes the size and shape of the parent rock masses, the minerals originally present remain intact. With chemical reaction, on the other hand, comes a change in the very composition of the minerals themselves, as well as a change in the physical state due to the formation of new compounds having their own physical properties.

Chemical weathering, therefore, is a much more complicated process than physical weathering. The chemical decay of minerals depends in large part on the fact that the surface environment contains the active chemical agents oxygen, carbon dioxide, and water vapour, which are rarely present in deep-seated environments. These three chemical agents tend to produce *oxides, carbonates*, and *hydrated compounds*, respectively, by combining with the primary minerals of the rocks that originated far below the earth's surface.

The low pressures and low temperatures of the surface environment, then, demand oxidized, carbonated; and hydrated minerals for stability. Among the primary minerals thus far mentioned, quartz is already an oxide, and it is indifferent to chemical combination with either carbon dioxide or water. Consequently it remains practically unchanged under surface conditions.

Not so the minerals Of the silicate groups. The ferromagnesians, for example, may contain only partially oxidized iron, with the result that decomposition of the molecules sets in during the further oxidation of the iron by atmospheric oxygen. During this decomposition the rest of the silicate molecule may combine chemically with water and thus become hydrated.

Just as the ferromagnesians decompose in the presence of atmospheric agents, so the feldspars may suffer chemical decay. The calcium oxide may be released from the silicate molecule and may unite with carbon dioxide to form a very common carbonate called calcite.

Similarly the sodium and potassium may be released to form various soluble salts. Meanwhile, the silicate part of the feldspar molecule Becomes hydrated by combination with water, much as the ferromagnesians do. The net result of this hydration and decomposition of the silicate molecules is a mixture of débris that we call clay. Since clay may consist of several minerals, depending on the composition of the original material, we apply the term clay minerals to the group as a whole.

We have introduced two new mineral names here. Let us pause a moment and look at them. We said the clay minerals are hydrated silicates. That means they still are silicates, but now water has become an essential part of the molecule. Here is the formula of one of them: $H_4Al_2Si_2O_9$, or $Al_2O_3 \cdot 2SiO_2 \cdot 2H_2O$, commonly called *kaolin.*

Other clay minerals may contain magnesium. Clays are composed of very small particles; and, as everyone knows, they are adhesive and slippery when wet. In the formation of the clay minerals, then, the hard and firm silicate minerals break down to finer débris. This is the important point to understand.

The other mineral, calcite, is calcium carbonate, $CaCO_3$. It almost invariably results from the decomposition of calcium bearing primary minerals. In some cases calcite is found in crystals that resemble somewhat the canine teeth of dogs, from which the fanciful name *dogtooth spar* arose. In comparison with most other minerals, calcite is relatively soluble in water, and especially so in the presence of carbon dioxide.

Hence, under many conditions of weathering the calcite may be carried away in solution by waters percolating through the ground. Quite similarly the soluble salts that are formed from the sodium and potassium in the silicates are carried away in solution and ultimately contribute to the salt supply of the oceans.

The whole process of weathering is thus an approach toward equilibrium or balance under surface conditions. With the exception of quartz, the primary minerals largely break down to new minerals (called secondary minerals because they are derived from the others) which are in adjustment with their environment of low pressures, low temperatures, and the chemically active gases of the atmosphere. Plate

4 illustrates some of the features of the secondary minerals. The careful reader may, at this point, recall that began with some emphasis on the slowness of the weathering process.

We may now summarize the process of chemical weathering something like this:

PRIMARY MINERALS	WEATHERING ENVIRONMENT	SECONDARY MINERALS
Quartz		Quartz
		Clay minerals
Feldspars	→ (By oxidation, carbonation, and hydration)	Calcite
Ferromagnesians		Salts of the sea

If it is so slow, how do geologists know the exact stages through which the rocks pass during their decay? Simply by being able to observe rocks in all stages of weathering, and from this evidence building up a connected sequence of events.

At the risk of becoming a bit technical, we shall let the reader look behind the curtains and see the methods by which the stages of weathering are studied.

Suppose we collect a sample of a partially weathered deep-seated rock from near the surface of a quarry, and then take a series of samples farther and farther down into the fresh, unaltered rock. We grind down chips of the rock with an abrasive until they are only a fraction of a hundredth of an inch in thickness.

These *thin sections* are then examined under a microscope. Let us start at the bottom, and see what the fresh, original rock looks like. For simplicity's sake we shall concern ourselves only with the feldspar minerals, which quite adequately illustrate the process.

Here we shall have to turn to Plate 4 and look at the photomicrographs toward the bottom. In the circle to the left, notice the large central crystal with its alternate white and dark bands, which fresh, unaltered feldspar. Notice that it is crystal clear. Now look at the middle picture. See how the upper crystal of feldspar is blotched and mottled, although the light and dark bands (narrower in this case) may still be seen.

The blotches are incipient changes, in which the feldspar is beginning to show the effects of surface conditions. When we ex- amine the photograph on the right, we see a much later stage of weathering in a sample taken from nearer the surface. In this thin section only a "ghost" outline of the feldspar crystal remains.

The original composition has almost wholly changed to clay minerals, although a few suggestions of the bands may still be seen. If we were to illustrate the fourth and final stage, it would be a handful of crumbly soil from the surface, where the originally crystalline rock has broken down to heterogeneous débris.

Not alone are there physical and chemical agents of weathering, but many biological agencies as well. Lichens, for example, cling to the surface of rocks and gradually dissolve some of their constituents for plant food.

These attacks commonly produce dull and pitted surfaces where the plant acids have leached the rock. In addition to this solvent action, plants may also exert mechanical forces by their growing roots. It is not uncommon to find trees growing from cracks and joints in rocks, where their enlarging roots have forced apart the rock by a type of wedge action.

Animals also contribute to the ultimate decay of rocks in several ways. Burrowing animals dig up buried material from the mantle rock and expose it to the more concentrated action of the atmosphere. In this connection most persons are familiar with Darwin's classical researches on earthworms, wherein he showed that in many places annually the earthworms on each acre of land ingest and cast aside 10 tons of soil, constantly exposing fresh material at the surface.

All of these factors contribute in some degree to the weathering process and afford one of the many examples of the close integration of geology with the biological as well as the physical sciences.

One of the most important factors in rock-weathering is the climate. In arid regions physical changes may predominate and chemical reactions be relatively inconsequential. Oxidation of the disintegrated rock, however, is fairly common, although hydration is relatively rare. Hence, in the formation of iron oxides the red ferric oxide may predominate.

The absence of reducing agents such as organic matter (humus) tends to keep the iron fully oxidized, so that red and brown colors among ancient rocks (such rocks are called red beds) are regarded by many as an important feature in identifying desert conditions of the geological past.

In humid climates chemical weathering predominates, because the active atmospheric gases are dissolved in the rain water that enters and seeps through the rocks. Here hydration is pronounced, and the net result is a thoroughly decayed rock. Temperature also plays a role in this case, because the speed of chemical reactions increases rapidly with slight rises of temperature. Hence in humid tropical regions weathering proceeds most rapidly, produces the most highly weathered débris, and consequently very thick soils.

Once the surface becomes covered with a mantle of decayed rock, however, the percolating rain water has to circulate farther to reach fresh surfaces, and hence its dissolved gases will be less concentrated. Furthermore, the cover of mantle rock protects the fresh rock below from sudden or extreme temperature changes, so that the physical forces become less effective.

Thus arises the commonly observed situation of a zone of mantle rock, with its surficial covering of *soil*, blanketing the fresh rock below, and with the degree of weathering becoming less pronounced downward. Some of the factors involved in the formations of soils and mantle rock are illustrated in Plate 5.

Fig. Bedrock Grades Upward Rock and Soil

If the weathered products accumulate indefinitely, weathering of the underlying rock virtually ceases. into mantle Other forces, however, are continually at work preventing this situation from developing in all but a few favored localities. Usually there is a fairly constant removal of the surface materials by gravitative pull down slopes, by winds, and by running water.

The surface of the mantle rock is thus gradually removed; and as the mantle thins, weathering processes are again able to penetrate more deeply. In this way there tends to develop a balance between the rate of weathering and the rate at which surface agents carry away the débris.

In the final analysis the removal of the weathered rock débris depends on the force of gravity. It is the component of gravity along the slope that determines the flow of water in streams, and winds are generated by relative gravitative effects between warm and cool bodies of air. There is one mode of transportation, however, that depends directly on gravitative pull and which involves no moving medium with its frictional grip on the débris.

We have already touched upon it: it is the downward tumble of loose blocks of rock from steep slopes. When rocks crack and break during weathering, fragments result.

If this disruption occurs on cliffs or steep slopes, the loosened débris immediately falls or rolls down, and collects as talus at the bottom. The slope of the talus is adjusted to the *angle of rest* of the blocks; it is the steepest angle at which the blocks may lie. Further breaking of the blocks in the talus pile causes shifting of the material, so that over long periods of time there is a gradual movement down slope.

A somewhat similar phenomenon takes place along river valleys, where the mantle rock on the valley walls gradually moves downward under gravitative pull. This slow movement is called *slope creep*. It is aided by the lubricating effect of water which percolates among the soil particles.

In some situations, such as in steep mountainous regions, the entire mountain side may give way and relatively rapidly slide down slope. Such landslides are often destructive. In these movements underground water also may loosen or lubricate the material, so that the force of gravity may more readily set it in motion.

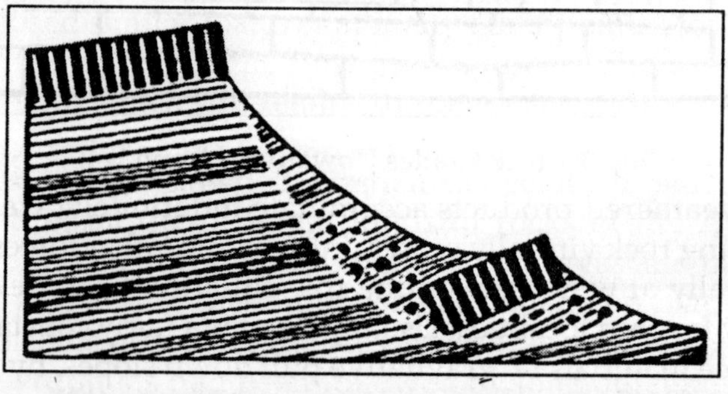

Fig. Diagram of a Landslide, Showing Displaced Massive Layer

These last two cases of movement are somewhat transitional between gravity alone as the operating force and the movements of surficial agents, like wind and water, which transport the débris directly.

In fact, most movement of rock débris is carried on by the geological agents, which are wind, running water (rivers and streams), glacial ice, shore agents (waves and currents along shores), and ground water, which seeps and percolates through the body of the rocks.

The current weathering breaks down the massive rocks into

smaller débris, which is thus subject to movement by the geological agents. We may accordingly turn our attention to these agents, considering first some of the underlying principles of transportation that are common to several of them.

WINDS AND TURBULENT MOTION

A modern, open-type freight car is about 50 ft. long and has a capacity of 50 tons. To transport 400,000,000 tons of gravel, sand, and mud requires 8,000,000 such cars, enough to make a train about 80,000 mi. long. Such a train would girdle the earth more than three times. And think of the freight bill!

It is estimated on good grounds that annually the Mississippi River carries enough sediment to fill such a train of cars. Yet the Mississippi is only one of multitudinous rivers on the earth; how much more, then, must be their total load! Add to that total the dust carried annually by the winds, and the great quantities of sediment shifted along the seashores, and the imagination reels under the resulting figures. There is no escaping the fact that this old globe is in the transportation business in a big way.

All of this transported rock débris must come from somewhere on the earth's surface. Unfortunately, too much of it is derived from our plowed fields, where wind and water are stripping off good topsoil at an alarming rate.

The widespread interest in the prevention of destructive erosion of our farm lands attests the importance of the subject in national economy. It is pertinent, therefore, to consider this geologic process somewhat closely. Accordingly, we propose to develop in some detail the principles underlying the movement of rock débris, in order to see just how it enters into the complete geological picture.

Winds usually are not smooth flows of air over the surface, but rather they act as alternations of gusts, eddies, and whirls. The velocity of the wind also varies considerably from place to place and from time to time. In similar fashion, when water flows over a stream bed, the irregularities along the bottom, such as boulders and stones, give rise to diversions of the current.

The result is that intermingling threads of water, as it were, wind and whirl about in many directions, with, however, a net movement downstream. Such irregular movement in a fluid is called turbulent flow. The irregular surfaces over which fluids commonly must flow in nature, and their average ranges of velocity, account for the fact that, except under rather quiet conditions, most movements of air and water are turbulent.

But in contrast to this irregular motion or turbulent flow, there is a quiet type called laminar motion which is also of some importance in geology.

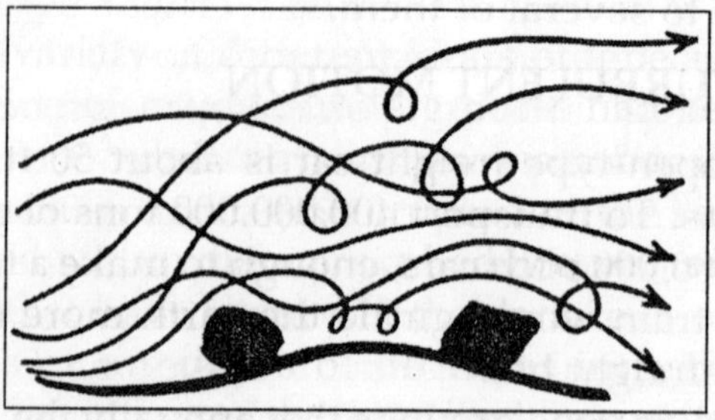

Fig. Eddies and Intermingling Threads of Liquid in Turbulent Flow

Turbulent motion in air and water, then, involves certain upward and sideward currents, as well as the forward mass movement. The irregularities of movement, especially those with upward components, may lift objects of small size up into the main stream, and so carry them along.

The greater the turbulence, broadly speaking, the greater is the lifting effect on the particles. The moment the particle is lifted off the ground into the liquid, however, it tends to settle again. The settling rate, or terminal velocity,* of the particle depends on its size, shape, density, and the nature of the fluid.

To consider all these factors in detail would require most of the remainder of this book. Accordingly, we shall content ourselves with those whose effects may readily be seen; the more complicated ones involve technicalities beyond the scope of this volume.

Most everyone would predict that if a sphere and a cube of the same material and volume are placed on a smooth level table, the sphere could more easily be set in motion by blowing on it than the cube. Furthermore, it takes no physicist to see that this result depends partly on the fact that the cube rests on the table with a greater part of its area than the sphere, Applying the same principle to a mixture of rock particles *of a given volume* but of different shapes, we should expeet that a gentle breeze will roll away the more spherical grains and leave the others, unless, indeed, the wind is strong enough to move them all.

This action of choosing certain grains on the basis of size or shape is called selective transportation, and it is an important factor in the work of wind and water. In fact, when a gentle wind acts on the dried sand of the seashore, it actually does roll the more spherical grains along with it, and leaves the flatter ones behind, This accounts for a phenomenon that you yourself may observe with a low-power magnifying glass, namely, that in general the sand grains in sand dunes are more rounded than those on the beach.

For a still better understanding of this situation, let us distribute a layer of sand grains of various sizes and shapes on a table-top and turn an electric fan on them. Most commonly the immediate effect will be a whirling cloud and a layer of débris over all the furniture; but if the fan is adjusted to move all but the largest grains, the selective action of the air current may be observed.

In the somewhat complex shifting of grains that results, it will be possible at least to distinguish that the very small grains of all shapes are whisked up into the air, the larger spherical grains are rolled along the surface of the table, and the largest particles remain behind. The grains that are rolled along constitute the traction load, the particles that are carried into the air constitute the suspension load, and the material left behind is the lag sediment.

The traction load, being confined to the surface of the ground, may be stopped by any obstacle in its path; whereas the suspension load, carried higher up, may move over the obstruction.

When later we analyse the factors causing deposition, we shall see that the deposits resulting from the action of wind or water are pretty intimately bound up with this difference in the traction and suspension loads.

But for the moment let us look more closely at the suspension load. We have seen that it is made up of a mixture of very small particles of all shapes, ranging from flat to spherical.

As soon as these rise above the surface, they are acted upon by gravity and tend to settle. Now there is a property of spheres that is unique among solids, whatever their shape, namely, a sphere has the minimum surface area per unit of volume.

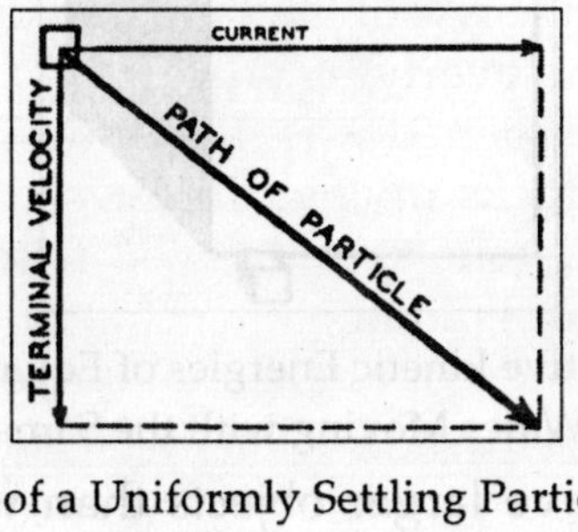

Fig. Path of a Uniformly Settling Particle Acted on by Forward Motion of a Fluid

Of particles having the same volume, then, the flat ones will encounter more resistance from the air because of their relatively greater areas, and so they will tend to settle more slowly than the spherical grains. As long as the lifting effects of the air currents are greater than the settling velocities of the grains, the grains will rise.

But there are so many intermingling currents in turbulent air that, as the grains rise, they are shifted about in a most complex manner. If they are shifted from an upward current to one having a horizontal movement, the grains will immediately start to fall.

In general they will not fall back to the starting-point, however, because as they settle the wind carries them forward. The path they follow, therefore, is a line that trends forward and downward simultaneously.

If we now consider the entire picture, we see that the action of the wind has been to sort out and classify the miscellaneous mixture of débris that results from rock-weathering. The various loads that are carried away, as well as the material left behind, have certain characteristics in terms of size, shape, and density.

The density, in turn, depends on the chemical or mineralogical composition of the grains; so it is apparent that many factors are involved. Because of the ways in which rock débris is sorted during transportation, we may expect that, if the resulting deposits are studied in detail, we should be able to learn something about the transporting agent. What we have been saying about wind applies equally well to running water, for in general the principles of transportation are the same for the two agents. We may note, however, that the differences in the densities of the two fluids partly determine their carrying-capacities.

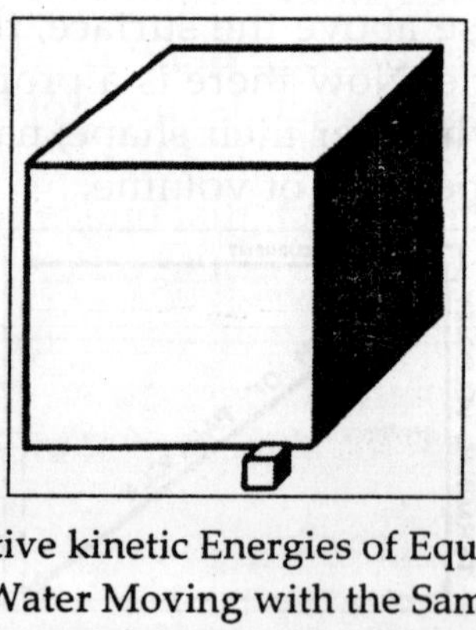

Fig. Comparative kinetic Energies of Equal Volumes of Wind and Water Moving with the Same Velocity.

That water can move larger objects than wind is not surprising when we remember that the density of air is about one one thousandth that of water, or, in other words, it is only one one-thousandth as heavy, volume for volume. Consequently, if equal volumes of air and water are moving with the same velocity, the kinetic energy, on which the effective work depends, is a thousand times as great in the water as in the air.

In fact, the average range of wind velocities, combined with its low density, pretty well confines wind action to the smaller particles like dust and sand. On the other hand, water suffers no such restrictions; and indeed it is capable of moving very large boulders.

Winds and running water are the most obvious transporting agents that display turbulence, but they are not the only ones. Ocean waves, as a matter of fact, also are turbulent as they dash against the shore within the line of white-capped breakers.

Waves are peculiar in one respect, however, in that they display a *pusating effect,* in a degree much more pronounced than in the general forward movement of wind and running water. We shall accordingly defer our treatment of ocean waves to a later point, especially since certain aspects of wave work also involve laminar motion.

Let us now look somewhat more closely at the movements of wind and see what sort of work it performs on weathered débris. It requires no great knowledge of geology to understand that wind work is most effective in arid regions.

There the decayed rock at the surface is usually quite dry and powdery, and it has no protective cover of vegetation. In desert regions, then, wind may be one of the most effective agents of transportation, and during windstorms may carry loads of sand and dust far and wide.

Not only deserts, but even our western plains, have witnessed the havoc and destruction of wind action. The plowed fields of those western states, where rainfall is scarcely 10 in. a year, are peculiarly susceptible to wind work.

The years 1995 and 1995 saw dust storms of such magnitude that practically complete destruction of large areas of farm land followed in their wake. The great clouds of dust were borne upward into the atmosphere and carried as far as the Atlantic seaboard, where they constituted dust storms of no mean magnitude.

There is more to wind work than we have mentioned. Not only is the sand and dust swirled about by the wind, but as it moves it may be blown against other rocks and act on them like a sand blast. In fact, wind abrasion, as this scouring effect is called, is of considerable importance in the arid Southwest; many of the prominent cliffs of varicolored sandstones there have been shaped and polished largely by wind abrasion. Commonly the traction load of sand, blown along near the ground, may also undermine the rocks and develop curious pedestal effects.

Among the various rocks that may readily be acted on by wind abrasion are sandstones. Composed, as they are, of sand grains loosely cemented together, individual grains are easily torn off by the blast of wind-borne débris, and they themselves join the load carried on by the wind.

Their erosional effectiveness is found, of course, in the fact that they lie within the range of sizes that can be most easily swept along by winds. As wind abrasion goes on, the rocks affected are worn away, so that the landscape is gradually changed. Plate 6 illustrates some of the aspects of wind as a geological agent.

Naturally enough, the material carried by wind tends ultimately to be deposited somewhere as the winds die down or as obstacles are met in the path of the traction load.

The deposits thus built up often assume characteristic shapes, and are composed of typical sizes of débris. Ordinary sand dunes, familiar to everyone, are wind deposits.

In their formation, once again the selective effect of turbulent agents enters the picture, and we shall have to devote some space to these depositional effects after we have considered a few other transportational agents.

The similarities between wind work and the work of running water are due mainly to similarities in their movement. In detail there are wide differences between them, because, whereas winds sweep over large areas simultaneously, rivers are confined to their valleys and concentrate their work along narrow sinuous paths. We may with profit, then, turn our attention to running water as a geological agent and see how and why it acts as it does. In this connection, Plate 7 should prove instructive.

Chapter 7

Valleys and Streams

Except in the most arid parts of the country, it is impossible to drive a car many miles without crossing bridges. The observant thus tend to recognize that there are several kinds of streams, some rushing through narrow rock gorges, others winding across wide flats with gentle valley walls in the distance; all of them, however, are confined to some kind of valley or trough cut below the general level of the land. Furthermore, everyone knows that main trunk rivers have tributaries, and these in turn have smaller branches, so that the whole drainage pattern, as spread out on a map, bears some resemblance to the branches of a tree. Finally, we all know that major rivers eventually flow into the oceans.

If we start at any point along a river and follow it upstream, we note as we pass successive tributary-mouths that the main river becomes a little smaller. Finally, near the headwaters the river itself has become a small stream, in some cases beginning in a swampy tract, in others tumbling down a mountain side as a brook, or merely heading in a dry ravine among the hills.

There are other features to be noted too. The valley flat, or flood plain, becomes narrower as it is followed upstream, until eventually it disappears and the stream occupies all of the valley bottom. The valley walls then become steeper, and the slope of the valley bottom itself increases toward the head.

Now let us see what happens at the very head of the valley during a rainstorm. Let s choose a simple case, where the terminating ravine is cut into mantle rock. When rain falls on land, it first soaks down through the spaces between the soil particles until the surface is thoroughly wetted. Then, as more rain falls, the water tends to collect in irregularities and to form tiny streamlets which flow downslope.

As the volume of water increases, it gains energy and sweeps along the finer soil particles in its path. By this process it carves a little trough, which may assume the proportions of a fair-sized gully if the land slopes away abruptly. Nearly everyone has seen such gullies along the sides of hills, especially where the soil is reasonably free from a protective covering of vegetation.

The rain water and its toad of débris are led by the slope into the ravine at the head of our valley. Here, during rainy weather, the concentration of water may become considerable; and as it tumbles along it carries mud, sand, and even pebbles along with it. When the rain has ceased, the ravine again becomes dry; but in the meantime several important things have happened.

In the first place, the rain water has washed rock débris down the slopes into the ravine, and the water there has carried much of this débris along, to supply it to the permanent stream farther down the valley. As a consequence of the removal of this material, the ravine gnaws a little farther into the hillside. Thus the valley is lengthened, and this lengthening always takes place in a headward direction.

Now, during any one rainstorm the amount of material swept from a square yard of land surface may not be great; but, given enough time, it becomes obvious that the valley will increase both in length and depth. Furthermore, the gully that we saw develop will itself increase by headward growth, and the rainfall that enters it will flow into the main ravine.

Thus the gully becomes a lengthened tributary of the main stream and will in time develop branches of its own. The reader at this point may well ask where this process will end. In brief, we may reply that the typical stream pattern is gradually enlarged by this headward growth until it comes into direct competition with another drainage system working in the opposite direction.

Each of these headward-cutting stream systems will then control a series of slopes leading to itself, with the result that a divide will lie between the opposing ravines. At this point a state of equilibrium is reached between the two drainage basins, and successive rainstorms will lower the divide, but, except under unusual conditions, the two drainage basins will remain separate.

In the foregoing discussion we assumed a simple case in which a supply of weathered rock was always available for the stream. This situation, however, is by no means universal. When rain falls on a bare slope of solid rock the clear rain water merely flows over the surface and does hot develop a gully. For water must have tools with which to work before it can attack solid rocks. In the first situation there was a layer of soil or decayed rock débris which was fine enough to be picked up by the running water.

Then carrying away of this débris must, of necessity, leave a hollow in the soil; and since water in rivulets flows as a continuous stream, the hollows must be shaped like the path of the water, which is sinuous and winds down the slope. If water which carries sand and pebbles with if now flows over solid rock, the particles strike against the rock surface and abrade it like sandpaper. Thus are other grains torn out; and these in turn act on more rock, gradually carving from even the most resistant rock a path for the water to follow.

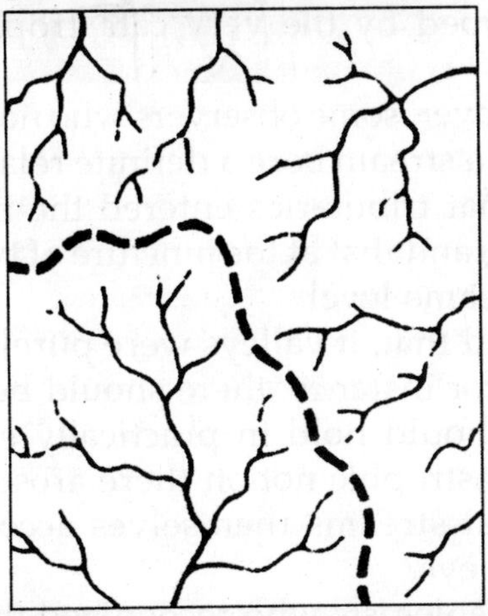

Fig. The Line of Dashes Locates the Divide Between opposing drainage basins

This wearing-away of rock masses by running water, or by air or ice, for that matter, equipped with tools for abrasion, is an important part of erosion, or the wearing-away of land masses by transporting agents. The carrying-along of weathered rock débris (and, in fact, the process of weathering itself) is also a part of erosion, since it tends to Wear away the land surface; and the tools thus furnished to the stream are in turn able to cut away the solid, and as yet unweathered, rocks that they encounter.

Observations of streams and their activities have been carried on in many parts of the world, and there is no doubt that the network of rivers everywhere is the result of just such headward encroachments of streams, spread out over great lengths of time.

We are able, then, to state with certainty that streams serve the function of delivering the rain water back to the oceans, and at the same time they transfer tremendous loads of débris from place to place over the earth. Eventually, all débris carried by the streams is dumped into the sea; and, given enough time under stable conditions, land areas may be worn down nearly to sea-level by this process of erosion.

These almost boringly simple principles of stream development have not always been recognized. Indeed, not much more than a century ago a controversy raged over even the manner in which valleys originate. It seemed unthinkable to many that streams could themselves have carved their valleys from the land.

A series of catastrophic events was called upon, during which great clefts opened in the earth, which later became occupied by their present streams. Even the source of water from rainfall was not recognized. Writers spoke of great subterranean chambers which fed the streams and which were tapped by the very catastrophes that formed the valleys.

These were, however, some observers who noted that in practically every case the size of a stream bore a definite relation to the size of the valley it occupied, that tributaries entered the main stream with an acute upstream angle, and that at the juncture of two streams the water generally met at the same level.

To them it seemed that, if valleys were purely accidental features due to earthquakes, for instance, there should be no reason why this systematic relation should hold in practically every case. Hence in opposition to the catastrophic notion there arose the present, almost self-evident, idea that streams themselves accomplish the work of cutting their own valleys.

One of the first geologists who recognized that valleys and other land features are due to the slow work of everyday forces, was James Hutton of Edinburgh. In 1795 he supported his views in a two-volume work on a Theory of the Earth, but unfortunately his style of writing was involved and his ideas did not gain the audience they deserved. Consequently, his friend, John Play fair, became his disciple and set out to explain Hutton's ideas "in a manner more popular and perspicuous than is done in his own writings."

But Play fair was no mere Boswell to his Dr. Johnson; rather he was a Huxley to his Darwin. He himself gave numerous examples of the principles announced by Hutton and attracted a respectable group of followers. Opposition remained strong, but with the accumulation of more and more data the fundamental truth of the new doctrines became clear. Hutton's observations were admirably summed up by himself in his now classic quotation, that "in the economy of the world I find no traces of a beginning,—no prospect of an end."

We are at present much farther along than Hutton or Playfair in out knowledge of the details of stream action; and among the several factors that can be evaluated, there are some that deserve our present attention.

All the factors that control the velocity of a stream along its channel are not easy to evaluate, because of the complexity of some of them. But we may at least discuss several of them that will help us to understand how a stream transports its load. Certainly, the *slope* of the bed is involved; and if is easy to see that the greater the slope the larger the component of gravity that operates on the water. The *discharge* of the stream, or the volume of water flowing through a given cross section per unit of time, is also involved: the greater the discharge, the higher the velocity.

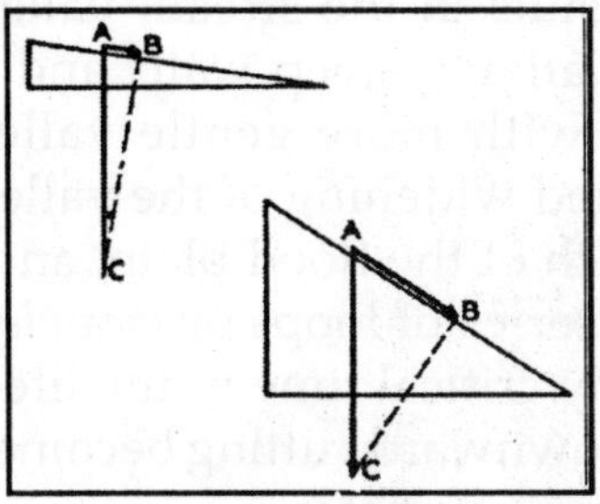

Fig. Effect of Gradient on Stream Velocity

Likewise the *form* of the channel, whether it is broad and shallow or narrow and deep, is an important factor because of the frictional effects of a large area of stream bed in contact with the water. The frictional effect is also dependent partly on *channel roughness,* for a smooth bed offers less friction to the flow of water than a rough one.

We may for out purposes, however, summarize these complexities as three main factors—the slope, the volume or discharge, and the frictional effects.

Whatever factors may predominate in a given case, the velocity of the water determines its kinetic energy, which in turn controls its carrying-power; and consequently the carrying-power is related to ail the factors we touched upon. As soon as material is picked up by the stream, some of the stream's energy is used in the transportation; and as a result the velocity of the stream is diminished.

The gradient of a stream is determined by the difference in altitude of the head and the mouth and by the distance between these two points. If a valley is incised into a plateau, where the relief is great, the gradient may be steeper than where the valley is cut into the walls of a rounded hill. The lowest level to which running water commonly may flow is sea level; and since gravitative forces seek to drain water to the lowest possible point, the waters of most streams ultimately find their way into the oceans.

The depth to which a stream may cut is determined in part by the original configuration of the land and by the difference in elevation between the head and mouth of the stream. There must inevitably come a time, then, when the stream has excavated its valley about as deeply as local conditions will permit. When this point is reached, the excess energy of the stream is expended in sideward cutting against its banks.

This sideward cutting may be due to original sinuosities of the channel, which divert the water from side to side as it flows along. The effect of the sideward swing is to widen the valley and to develop a flood plain along the bottom of the valley. Thus the change in direction of cutting results in the stream valley changing from a V-shaped notch, with relatively steep walls and a stoop gradient, to a flat U-shaped valley, with more gentle valley walls and a gentle gradient. With continued widening of the valley the stream is unable to cover the entire width of the flood plain, and hence it winds about the valley bottom in a series of loops or *meanders*.

Now, if there is any critical time in the life-history of a stream, it is this one, where the downward cutting becomes less pronounced and the sideward cutting predominates.

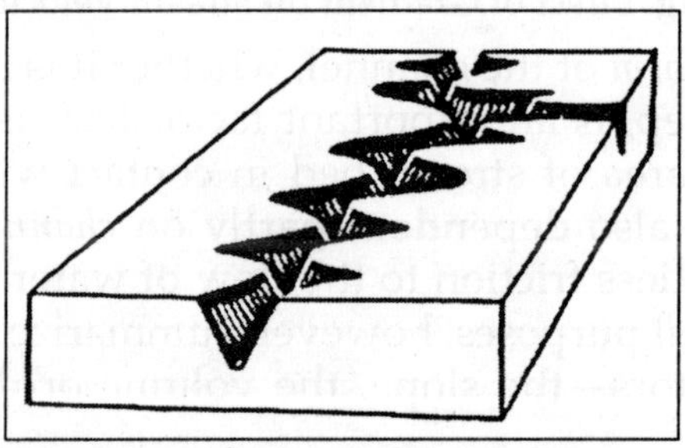

Fig. Block Diagram of a Youthful Valley, Showing its Typical V-Shaped Cross Section

The change in direction of cutting at this stage results in the stream valley changing its cross-sectional shape, as we have seen. The point at which the flood plain begins to develop is easily recognized.

The shift in the direction of dominant cutting is expressed by saying that the valley has passed from youth to maturity. The downward-cutting stage, with its steep gradients, steep walls, rapid movement, and small volume, marks the youthful stage of the stream's life-cycle.

The flood-plain stage, characteristic of its maturity, is marked by sideward cutting, a lessened gradient, and a greater volume of water. Finally, with a still greater width of the flood plain, an even lesser gradient, and slow movement, the old-age period of the stream is entered. The distinction between maturity and old age is one of degree rather than of kind, and it is not as sharply marked off as the change from youth to maturity. Refer, in this connection, to the photographs of streams and valleys in Plate 8.

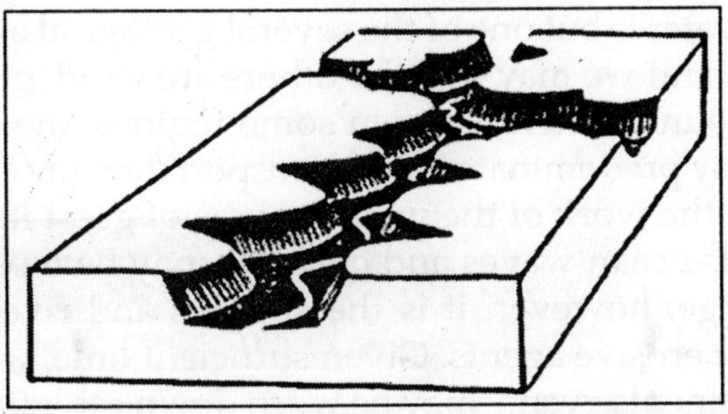

Fig. Block Diagram of a Mature Valley, with its Broad U-Shaped Cross Section, Due to the Presence of a Flood Plain

We must remember that we have here a dynamic process-that, while the stream is continually cutting headward in its youthful, lengthening ravines, the older parts of the valley are approaching the tower level to which they may cut, and that near the mouth the river may even be in the old-age stage.

This accounts for the fact that, as a river is followed downstream, the configuration of its valley changes markedly. IT IS important to recognize that this age classification of valleys is not based on time in years, but rather on the characteristics

Fig. Block Diagram of a Valley in Old Age

The stream meanders at will over the broad good plain of the valley itself in terms of the stages through which it passes. In fact, the length of time required for such a set of events depends upon the nature of the rocks, on the climate, the relief of the region, and so on, so that some streams may pass through their entire cycles while others are slowly passing through maturity.

The great importance of streams from an earth-science point of view is that they act as agents in wearing down the land and in carrying rock débris from higher to lower elevations. Ultimately, if enough time is afforded them, this débris is carried to the seas and oceans, and the general level of the land is lowered until only flat featureless plains remain.

Running water is but one of the several geological agents that wear away the land, and we may that the others are wind, glacial ice, shore agents, and ground water. Now, in some regions one or the other of these agents may predominate in the transportation of rock débris: thus, in arid regions the work of the wind may be of great importance, and along the coasts ocean waves and currents may be most effective.

By and large, however, it is the streams and rivers that are the most important erosive agents. Given sufficient time, whole regions—nay, entire continents even—may be worn down essentially to sea level by the relentless gradational activities of streams. Before we consider some of the other agents, therefore, we may complete out picture of running water by turning now to the *regional*, as opposed to the *single-valley*, aspects of running water and its geological work.

THE EVERLASTING HILLS

It is difficult to conceive of natural processes so slow that during the entire course of written history they have effected no outstanding changes; and yet our terrestrial globe knows many such processes slowly but inexorably altering its surface configuration through the ages.

As far back as reliable records go, the general configuration of land and sea has been about the same, and the ruins of the most ancient civilizations lie in geographic settings not markedly different from those when pomp and ceremony graced their age-old halls.

True, everyone realizes that some changes have taken place. Earthquakes, volcanic outbursts, tidal waves—all have left their marks here and there on the earth's surface.

By and large, however, the same valleys are still present, the same mountain ranges, the same seas and oceans. We today use the same passes over the Alps that were traversed by Hannibal, and the same majestic glaciers wind their frozen way along the same rock-ribbed clefts. Little wonder, then, that as far as average human experience goes, the hills are considered everlasting.

Notwithstanding this apparent permanence of hills and mountains, Tennyson was right.

The hills *are* shadows. Every trickling rivulet, every gust of wind, carries with it some rock débris. Therefore, given time enough, whole land areas will be carried away by wind and water, each agent constantly at work reducing the continents toward sea level.

True, it may require millions of years; but after all, a million years is not so very long in comparison with the awe-inspiring length of geological time. For definite evidence, which we shall examine later, indicates that the age of the earth is to be reckoned in billions of years. Even the most case-hardened skeptic will grant that lots of things can happen in a billion years, though it is doubtful if any but the most imaginative can grasp the real meaning of such a great lapse of time.

If the process of wearing away the land is so slow from a human viewpoint, how do we know that any vital changes do take place?

Observations by geologists in many lands have shown that, from an erosional point of view, there are areas in all stages of development; and by classifying these areas we are able to discern the underlying pattern of regional erosion.

Thus it is possible to work out a sequence of events that not only explains the land forms themselves but actually permits the development of a logical theory of the origin of land forms, relating each feature to the geological processes that developed it.

You may remember that we anticipated the present discussion when we described how streams develop first a small valley in which down cutting is dominant, followed later by a dominance of sideward cutting which widens out the valley and forms a flood plain. Furthermore, as the valley increases in length, it develops tributaries, until an entire area is covered by a network of streams.

We saw how all these features fitted hand in glove with the idea of a gradual development of stream valleys through several stages. .

In our first consideration of stream valleys we confined our attention to a single valley system; but of course there are many valley systems in any given area of a continent, and all of them are going through similar stages of development. Consequently, there should also be cycles of development through which entire areas go as a result of normal stream development.

The study of numerous regions does, in fact, permit us to pick out a sequence of regional forms. For example, there are large tracts of land in Utah and Arizona, called the Colorado Plateau, where the elevated land surface is nearly fiat between the gorges and canyons chat here and there deeply incise it.

In other regions, as in much of West Virginia, the land surface is mainly in slope, with very little level land, and with streams cutting into every hill and mountain.

Finally, we have examples of regions, such as the lower Mississippi Valley, in which the land surface is nearly flat. In such areas a winding stream, which is not confined within a deep gorge as in the first example, flows on a wide, shallow plain with very gentle valley walls.

Between these several types of surface are all gradations, which a detailed analysis shows result from a continuous process. We may sketch the broad outlines of regional development by considering some large plateau area, high above sea level, ignor

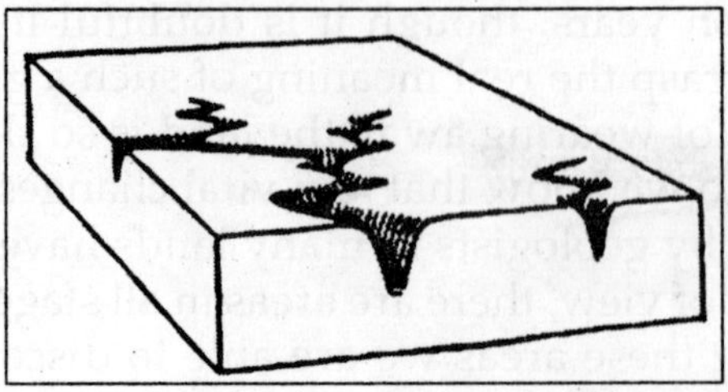

Fig. Block Diagram of a Youthful Region

Note the dominance of uplanding for the present the question of how the plateau originated. The elevation of the plateau gives steep gradients to streams; and as time goes on, we may visualize the slow but unceasing excavation that produces valleys in the plateau, starting at the edges and gradually lengthening toward the interior.

The valleys will be deep and steep sided because of the relative height of the plateau; and the tributaries will be similar to the main streams, but smaller. In the course of events, them we may picture to ourselves an area in which the flat top of the plateau still stretches out in unbroken monotony for many miles, but with here and there a localized drainage pattern made up of a main gorge and its tributaries, the gorges perhaps several thousand feet deep.

If we allow time enough for the main gorge to be cut down as far as the base of the plateau, it may even have a narrow strip of floodplain adjacent to the stream. Between the flat top of the plateau and the flat stretch of the flood plain there is considerable relief; and to distinguish the several levels we call the plateau top the upland, the valley walls the slope, and the floodplains the bottom land.

Let's stop a moment and see whether we can describe the area in this stage of development a little more precisely. We see before us a region in which most of the surface as seen on a map

The white area is upland. Note the narrow flood plain at the of the valley is still an unbroken upland. Here and there, where the streams have penetrated the plateau, there are steep slopes leading down to the valley bottoms. The total amount of slope, again considered as an area on a map, may be only a few per cent of the total.

Finally, along the bottoms of the larger canyons there are narrow flood plains here and there, making up the very small remainder of the total map area. Since we started out with 100 per cent upland, it is obvious that the present stage of development cannot be very far along, because there still is a great preponderance of upland left.

Consequently, we call this a youthful stage, on the basis of the ratios of upland to slope and upland to bottom land.

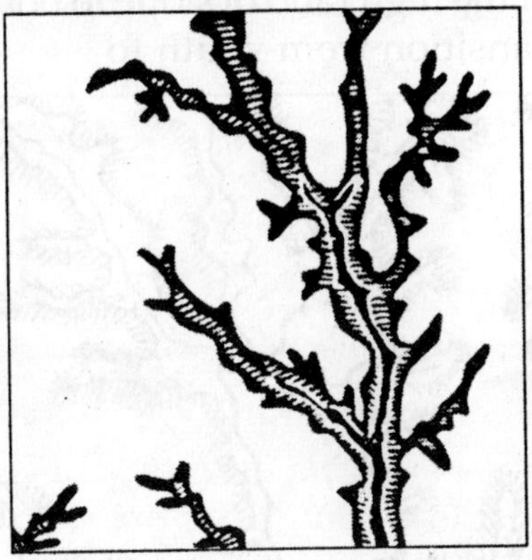

Fig. Map of Youthful Region

While we have been considering this first stage, the process of erosion has continued; and, coming back now a million or so years later, we find that the streams have grown headward farther and farther into the plateau and have spread a network of tributaries over most of the region. We may still find a few small areas, however, where the original flat top of the plateau remains; but, if so, these are now merely narrow, flat-topped divides between drainage basins.

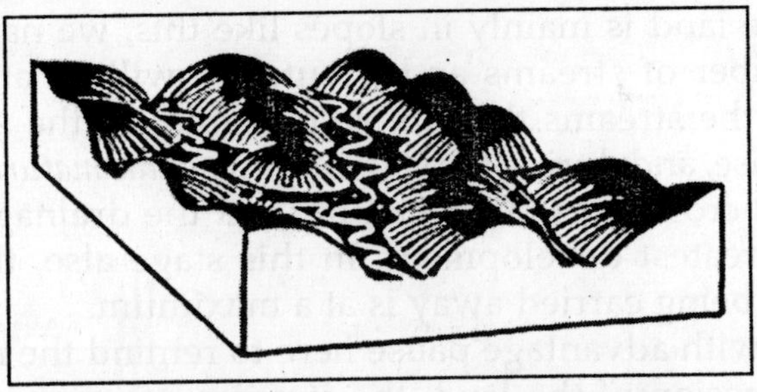

Fig. Block Diagram of a Mature Region.

With the increased amount of stream-cutting, the amount of land in slope has greatly increased, although the slopes themselves may not be quite as steep as they were in the earlier stages. The major streams, too, have developed wider flood plains, and as a consequence we find that the ratios of upland, slope, and bottom land have changed considerably. The upland has been decreased to a small percentage of the total map area or has been obliterated entirely. Meanwhile the area in slope has increased to dominance in the region, and the bottom land itself has grown to some extent in area.

Here the slopes clearly predominate, and the characteristic youthful aspects are gone. We accordingly describe the region as mature on the basis of the high ratio of slope to bottom land and upland. Somewhere in the transition from youth to

Fig. Map of Mature Region

The upland has entirely disappeared, and the area in flood plain (dotted) has increased maturity the last remnants of the original upland entirely disappear, so that slopes occupy all of the region except for the still relatively small amount of bottom land.

When the land is mainly in slopes like this, we naturally expect that the number of streams and tributaries will be at a maximum, because it is the streams that are responsible for the slopes. This is actually the case, and during the stage called *middle maturity* the streams are as closely crowded as they can be, and the drainage pattern has reached its greatest development. In this stage also, the amount of débris that is being carried away is at a maximum.

We may with advantage pause here to remind the reader that all during this erosion of the land, the streams are carrying away rock débris. The process we are describing is thus *unidirectional*: all the factors involved tend in one direction, the direction of *degradation* or down cutting.

During regional youth, when streams are few, the amount of débris carried off is relatively small; but as more streams incise the land, and as the proportion of slope in- creases, the total load carried away also increases, until in middle maturity the rate of removing material reaches a maximum, as we have said. Beyond middle maturity, we shall find that the area in slope decreases; and as it does, the amount of rock débris carried away also decreases.

While we are pausing, we may also draw an interesting contrast between the maturity of land areas and the maturity of individual valleys. A mature valley, we recall, is one in which down cutting has become negligible and sideward cutting dominant. Does it follow that a mature region will have mainly mature streams?

Not at all, because if the number of streams has been increasing all the time, it is obvious that there must have been considerable high land available so that new tributaries could continually cut headward. These new tributaries will be in their dominantly down cutting stage, so that, as a matter of fact, most of the smaller streams in a mature region are themselves youthful.

This fact ought not to cause confusion, however, as long as we remember that we are talking about two different things when we contrast stream age and regional age. Furthermore, as emphasized earlier, the ages are not measured in years, but in stages of development. According to the natural laws that control the development of both streams and of regions, it follows that during maturity in a regional sense the conditions are most suited to youthfulness in the majority of tributary streams.

Now, as we follow our regional development still further, the reader will be able to anticipate that, as the numerous streams of middle maturity continue their work, they will gradually deepen their valleys, and finally more and more will reach the lower level of down cutting. From this point on, sideward cutting will increase relatively, and the total amount of bottom land in the area will correspondingly increase. Consequently, a few large streams will take the place

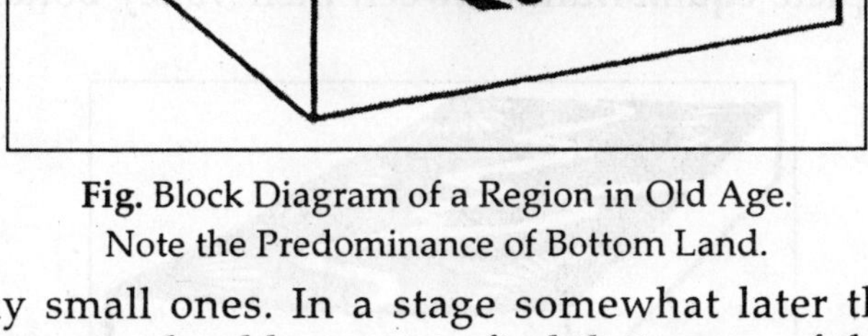

Fig. Block Diagram of a Region in Old Age. Note the Predominance of Bottom Land.

of many small ones. In a stage somewhat later than middle maturity, then, we should expect to find that many of these streams have relatively wide flood plains and that the slopes are no longer nearly as steep as they were. In fact, the divides between adjacent streams tend to become rounded off into gentle hills, and much of the ruggedness of the earlier stages is lost. The upland, we saw, disappeared long ago; and with the widening of the flood plains the ratio between slope and bottom land is increasing in favour of the latter.

The stage will inevitably be reached, given enough time, when the total area occupied by bottom land will be greater than that occupied by slopes.

When this situation develops, we say that the region has reached old age. These three principle stages of regional age are shown in the photographs of Plate 9.

Fig. Map of Region in Old Age. The Bottom Land Predominates

Here, then, are three distinct stages through which entire regions may pass; and as we should expect, there are all possible gradations between the stages, because this is a continuous process and moves along as smoothly as time itself, it will be interesting to venture a bit farther and inquire what the very final product of this whole process is well, as the streams continue carrying away the débris, the slopes become still more gentle, the divides are lowered still more in height, and the main streams become more and more sluggish as they reach a state of complete equilibrium between their valley bottoms and sea level.

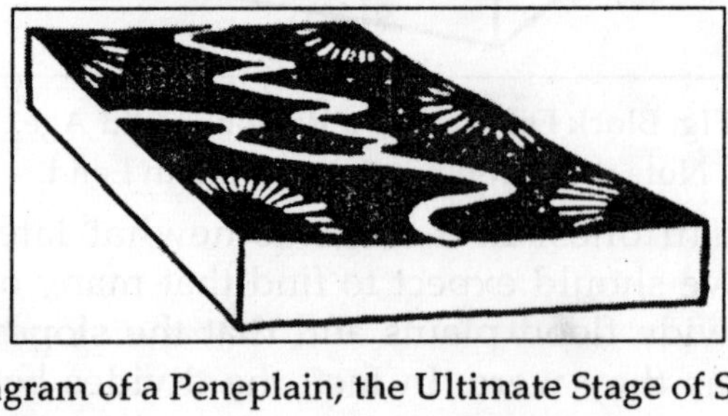

Fig. Block Diagram of a Peneplain; the Ultimate Stage of Stream Erosion

No stream can cut much below sea level, of course; and in addition, there must be a slight gradient from the ocean upward into the continent, to allow the water in the streams to flow into the sea. A flat, or nearly featureless, plain, with sluggish streams that slowly wind toward the seas?

All about your point of vantage stretches this monotonous surface, with the horizon broken only by a few low hills; it is, indeed, a nearly horizontal line. If you can picture such a plain, then you are visualizing the final peneplain stage of the cycle of erosion, the lowest level to which streams can reduce a land area if they are given unlimited time in which to work.

We realise that it is difficult to obtain such a long-time point of view on short notice, but we believe that the logical sequence of the events described should impress the reader with the essential simplicity of the process when it is shorn of details. There are complexities galore in actual practice.

The rocks of the area may not be all of the same kinds, so that some will yield more readily than others to weathering and erosion; the original laud may not have been a plateau at all; and finally, various other types of events may interrupt the process so that not enough time is afforded for a completion of all the stages.

Changes in the levels of land and sea may, and usually do, interrupt the procedure, so that one cycle of erosion becomes superimposed on another until such a degree of complexity is reached that even a full-fledged geologist finds great difficulty in deciphering the record.

Fortunately, there are many regions available for study in which the complexities are not insurmountable. It is from these that the underlying principles have been drawn; and, armed with the laws that control the situation, geologists are pretty well able to attack even the more special and complex situations. We may see, then, why we can state with reasonable assurance that land areas undergo systematic changes with time, because in nature we find regions in all stage f erosional development.

It is a monument to human thinking that, from the a confusing and chaotic abundance of land forms, it has h to set up a logical system of events that both explains a the many details of land sculpture. Furthermore relating these details to the underlying erosio finishing touch, as it were, on our ability t ways in which nature does her work.

The study of land forms, and

accordance with the processes that formed them, is called physiography, an important branch of earth science. In the discussion of this subject we have been concerned with the land forms that result from the erosive work of streams, but of course streams are only one of several geological agents.

Each of the agents—and here are included wind, glacial ice, the waves and currents of the oceans, and ground water—have their own sequences of land forms, due both to erosion and deposition. For we must not forget that while erosion takes place at one point, deposition takes place somewhere else. Thus far we have not said much about these depositional types of topography, but their time and place will come.

Finally, in addition to the particular work that each agent performs, usually more than one act simultaneously on a region, thus giving rise to a whole series of related cycles moving along side by side. As an example of such a situation we may think of most any high mountain region where wind, streams, and glaciers are all contributing to the down cutting process.

The past history of the earth is marked by numerous cases in which whole masses of land have been cut down, from lofty mountains to actual peneplains. The process we have been sketching is thus not purely hypothetical, or one that is just starting for the first time; but rather it is a process which has seen fruition on many occasions in the past. So important are these ancient peneplains in marking off stages of the earth's history that later we shall have occasion to mention them time and again.

Chapter 8

Underground Water and Laminar Flow

Athanasius Kircher, one of the greatest writers on natural science in the seventeenth century, thought that all rivers and streams arose from underground lakes or reservoirs, which in turn were supplied by conduits leading to them from the oceans. When we remember that this was back in 1665, before many scientific observations had been made, we see that Kircher's idea was rather ingenious.

He at least recognized that the lithosphere contained abundant water, and that this water somehow escaped to the surface, where it contributed to the supply in rivers and streams. In short, he sensed what we know quite well today, namely, that water is one of the most ubiquitous things on earth. The hydrosphere, which covers some three fourths of the globe's surface, is named after it; and the atmosphere contains on an average about I per cent of water vapour. The lithosphere also has its own important share of water.

Whenever rain falls on the ground, some of the water inevitably sinks down among the interstices of the rock and thus enters the body of the lithosphere. This is called ground water, and locally it may completely saturate the pores of the rocks. Ground water circulates beneath the surface, and during its travels it performs work which is geologically important. Suppose we accompany it for a part of its journey and see what happens.

We mentioned the pores in rocks. To the average person the very phrase porous rock" seems a contradiction, and it may be well to explain further. We already know that rocks are not the impregnable, everlasting things we may have thought they were; so it is not difficult to understand also that most rocks are not perfectly dense.

They are made up of solid particles like crystals and grains, among which are open spaces, so that a varying amount of pore space is present in almost every rock, at least near the surface of the earth. To make the notion concrete, imagine a rock made up of small spheres, packed together as closely as possible. Then there would still be about 26 per cent of open spaces among the spheres.

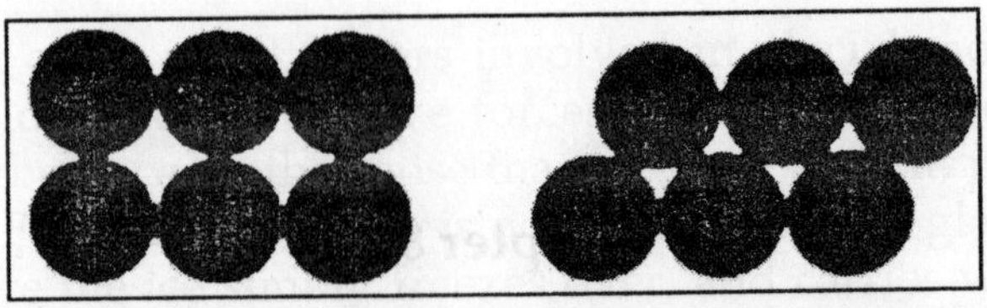

Fig. Spheres Packed in Two Manners

Here is an interesting thing, however: no matter what size the spheres are, any given volume of them packed in the same manner has exactly the same amount of pore space. That means that a cubic foot of rock made up of spheres one-hundredth of an inch in diameter would have as much water-holding capacity as a cubic foot of rock made up of spheres 1 in. or more in diameter. With more irregular grains, and also with mixtures of various sizes, this porosity of course varies considerably.

Rock porosity is obviously important in the general subject of ground water, but even more significant is the rate at which the water is able to circulate among the rock pores. Consider our 1-in. spheres, for example, in which the open spaces will be rather large. We can easily see that water would have little difficulty in moving about between such spheres. Permeability is a measure of the ease with which fluids move among the pores of rocks.

Hence a rock made up of these 1-in. spheres would be quite permeable. Now if the spheres are gradually decreased in size, the spaces between them also become smaller, and consequently the water must pass through smaller and smaller openings in the rock. Eventually a stage is reached in which the total surface area becomes so great that the spheres begin to attract an appreciable part of the water by molecular forces.

When this happens, it requires a greater pressure or a longer time, or both, for the water to move through the pore spaces. Permeability, then, plays an important role in the movement of ground water. Near the surface, of course, the rocks may be shot through with cracks and crevices, so that the water has many lines of descent; but with increasing depth, such openings are less common, and the water in large part must pass through the pores of the rocks. The latter may be likened to a sort of continuous sieve, with openings of various sizes.

As a general rule, ground water moves very slowly, because of the variations in the permeability of the rocks it encounters, and also because of the character of the forces which set it in motion. We shall discuss these forces later, but for the present perhaps we had better get back to the surface and see what is going on there.

Three things may happen to rain that falls on the land. It may evaporate and return to the atmosphere, it may run off down the slope, or it may sink into the ground. If it goes back into the atmosphere, it ultimately contributes to the rainfall somewhere else; if it runs downslope, it feeds into streams and so contributes to the work of running water.

If it sinks down into the earth, it becomes ground water, and so enters our present bailiwick, Now the relative proportions of rain water that follow one or another of these three paths clearly depend on the conditions of the locality. In a region of steep slopes, most of the rain flows directly to streams; but in a region dominantly fiat, a fairly large amount of water sinks downward into the ground. Atmospheric conditions, climate, and a number of other factors control the rate of evaporation.

Whatever proportion of rain water enters the soil and rocks beneath the surface tends to sinks; straight down under the influence of gravity until it reaches a zone where the rocks are already completely saturated. Everybody knows that the soil is not usually sopping wet to the surface, but that, when wells are dug, it is necessary to penetrate some distance downward before a supply of water is found.

The level below which the rocks are saturated is called the ground-water table, and in any given region it bears a certain relation to the configuration of the land surface above it. A considerable amount of fairly exact data has been collected on the subject of the ground-water table, particularly by observing the height of water in wells. From such data it is easy to construct a map or surface showing the relationship of the ground-water table to the land surface above.

The adjacent diagram shows the general relation of these two surfaces. Note that the ground-water table is at the same level as

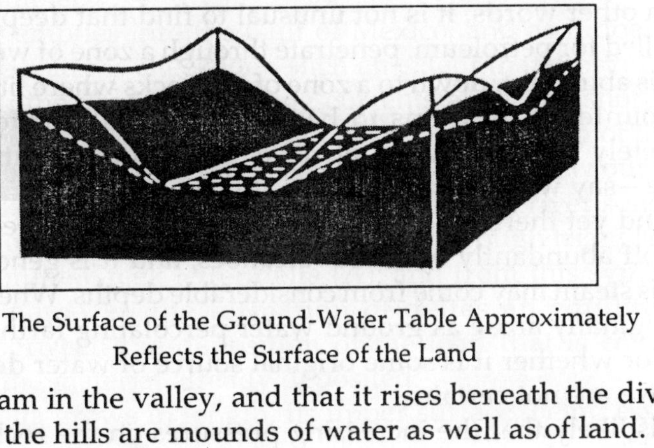

Fig. The Surface of the Ground-Water Table Approximately Reflects the Surface of the Land

The stream in the valley, and that it rises beneath the divides, so that literally the hills are mounds of water as well as of land. The fact that the ground-water level coincides with the stream level is important, because it explains why streams continue to flow even after the rain has ceased. The water that entered the ground during rains is slowly fed to the streams, and so maintains them in the dry weather.

This is not universally true, of course, because ravines and gullies usually are quite dry between rains. The reason, obviously, is that these little fellows have not yet cut down far enough to intersect the permanent ground-water table. As the ravine is followed downslope, however, a place is reached where it carries a small permanent stream. The head of this stream marks the level of the ground water.

We see from the foregoing discussion that where the groundwater table intersects the surface, water issues from the earth. Thus *springs* are nothing but seepages of ground water where the land surface is cut by the water table. Naturally enough, the water that seeps out usually comes from a higher level.

There are many complexities among individual springs, and we shall not pursue them all. In some cases the water issues from more permeable beds; at others there may be hydrostatic pressures involved; and so on. In the main, however, it is reasonably accurate to say that wherever the land surface dips down beneath the ground-water table, there water will issue from the earth. Lakes and swamps, then, are merely depressions in the land where the ground-water table comes out of concealment into the open.

The upper level of ground water can be clearly established. Let us see now whether there is also a lower level, or whether, conceivably, the ground water fills all rocks far down into the interior of the earth. Observation again comes to our aid, for we know that in many deep mines and deep wells the rocks are quite dry. Surprisingly enough, this is only a matter of some thousands of feet.

In other words, it is not unusual to find that deep wells, such as are drilled for petroleum, penetrate through a zone of wet rocks, where water is abundant, down to a zone of dry rocks where little or no water is encountered. It seems to be pretty generally true, then, that a completely saturated zone of rocks occurs only relatively near the surface—say within a mile or two at most.

And yet there is some water at greater depths, because steam is given off abundantly by most volcanoes, and it is generally believed that this steam may come from considerable depths. Whether it is water that originally arose as ground water percolating farther and farther down, or whether it is some original source of water deep within the earth, we are not certain.

It is likely that the increasing pressures on the rocks far beneath the surface may have an effect on the depth to which the ground water penetrates. Eventually, if one considers a point deep enough beneath the surface, the pressures of the overlying rocks are so great that none of the rocks buried there has open cavities or pores; but such a depth is a matter of miles rather than of feet.

One reason why the problem is not completely solved is that the number of deep wells that go below the ground water is limited. After all, most wells are drilled for water itself, and so they stop when they penetrate some short distance into the ground water.

Occasionally, a rather curious situation is met when wells are drilled. The water, instead of remaining at about the level of the ground-water table, rises up into the well, and in some cases flows out on the surface. Such wells are called artesian. The explanation of these wells is not difficult in the light of what we already know.

Fig. Conditions Necessary for an Artesian Well.

Consider a layer of highly permeable rock between two that are relatively non-permeable or impervious. If these rock layers are inclined downward from the surface, the ground water within the permeable bed is confined by the impervious layers and thus exerts a pressure against its surroundings. If man drills a well through the overlying impervious layer, he affords the confined water an opportunity of relieving its pressure, which is promptly done by some of the water flowing up through the well.

Artesian wells are very important, economically speaking. Many cities and towns depend entirely on artesian waters; and even in Chicago, with a huge take at its doorstep, many industrial firms have used artesian water. One of the important sources of this water is a sandstone which is exposed at the surface in Wisconsin at altitudes higher than Chicago.

These beds dip downward beneath the surface toward the south, affording a means of circulation of rain water from the more elevated exposed catchment area to the metropolis distant several hundred miles. But after all, artesian conditions demand special relations among the rocks.

More usually there are alternate layers of rock of varying degrees of porosity and permeability through which the ground water as a whole tends to circulate. When we come to consider this general circulation of ground water, we find ourselves faced with some problems, for it is difficult to make direct observations on the movement of water beneath the surface of the earth.

We do know definitely, however, that ground water moves very slowly. The water percolates among the rocks at a rate that may be expressed as a few hundred feet per year. This sluggishness is a result of the great frictional resistance the water encounters as it migrates among the small interstices in the rocks.

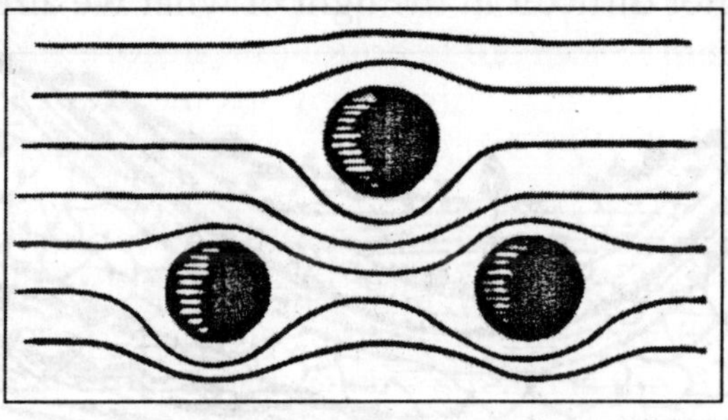

Fig. Streamlines in Laminar Flow

Ground water displays no turbulence in its slow movement among the pores. Here is a case where water flows with another type of movement, which is called laminar flow. * In laminar flow, any given thread of the liquid retains its identity and flows smoothly alongside its neighbors. It is a sort of streamline flow which adjusts itself to irregularities in its path by curving around them, rather than by setting up whirls and eddies in the current.

Laminar flow involves the concept of *viscosity*, which is an internal friction in the liquid, so that there is a drag of the liquid layers one on the other. Now, water actually is a viscous fluid; but the running water of surface streams, because of its relatively high velocity and the irregularities in its path, seldom displays laminar motion. Beneath the surface, however, laboratory experiments have indicated that the slow movement of water through rock pores due to gravity is definitely laminar.

Some readers may be interested in the experiments that study the flow of water and other fluids through rocks. In the field of petroleum geology it is important that the nature of the movement of oil and its associated water be known. Consequently, cores of actual rock are mounted in suitable apparatus and water or other liquids are forced through under pressure. The amount of water flowing through per unit time and the pressures that cause this flow are measured.

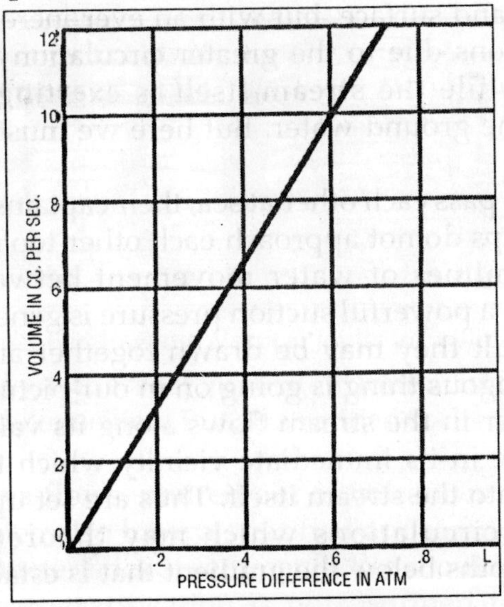

Fig. Graph Showing Flow of Water Through Cisco Sandstone.

These observations constitute the factual data. From theoretical grounds it is known that if the volume of water flowing through the rock is directly proportional to the difference in pressure at the two ends of the rock core, then the flow is laminar.

In other words, the graph of the experiment is a straight line in laminar flow; whereas, if the curve departs from a straight line, then turbulence is present, for the energy displayed in eddies and whirls reduces the volume below its expected theoretical amount. Graphs, such as the accompanying one, are based on experiments of this sort. Covering a wide range of pressures and velocities, they prove that the ordinary gravity movement of water in rocks is laminar.

To get some idea of the causes of ground-water circulation, we shall have to assume a rather simple situation in nature. Imagine a valley between two divides, in which it has been raining "for forty days and forty nights," as it were, until all the pore space in the rocks of the divides is filled with water. Now imagine the rain to cease. The excess water on the surface will flow down to the stream and be carried away. Meanwhile the surface of the ground water in the rock coincides with the surface of the land.

This underground water also has a gradient, then, down which it may flow. Gravity is able to operate on the mass of water, and a circulation results. Some of the water, perhaps a large proportion of it, tends to move down parallel to the land surface toward the stream. The slope of the ground-water surface will tend to remain about parallel to the land surface, but with an ever increasing departure in its higher portions due to the greater circulation from those higher points, Meanwhile the stream itself is exerting an effect on the circulation of the ground water. But here we must pause for a bit of theory.

If two ships pass each other at sea, their captains have to be careful so that their ships do not approach each other too closely. If they do, then the streamlines of water movement between the boats are constricted, and a powerful suction pressure is generated between the ships. As a result they may be drawn together and collide. Now a somewhat analogous thing is going on in our picture.

As the water in the stream flows along its valley, it generates a suction pressure in its immediate vicinity which tends to draw the ground water into the stream itself. Thus are set up slow and almost imperceptible circulations which may theoretically extend to considerable depths below the gradient that is established by gravity alone.

For practical purposes, however, effective circulation probably does not extend much below the level of the main stream in the region. Below some such level the slight mass movement that may be present is probably negligible.

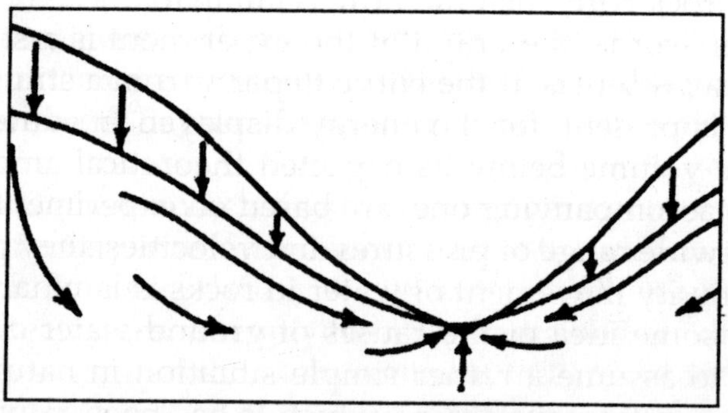

Fig. Ground Water Descends Vertically from the Surface of the Land to the Groundwater Table.

The very slow circulation of ground water, combined with the fact that it is passing through what may be considered a continuous sieve, means that most ground water does not transport solid matter. By far the greatest transportation work of ground water is done by solution, and consequently we should expect its action to be limited to the more soluble minerals. In fact, *limestone,* among all the common rocks, is most susceptible to ground-water attack, because the calcite of which it is composed is the most soluble of the commoner minerals.

When we talk about the solubility of calcite, we do not want our readers to think that it compares with the solubility of sugar, for example. As far as ordinary things go, calcite is relatively insoluble; but in contrast with most of the rocks that occur in nature, it is readily subject to solvent action by percolating waters. To make the thing more specific, we may say that only about 0.001 gm. of calcite can be dissolved in 100 gm. of carbon-dioxide-free water.

That isn't much in comparison with sugar, say, of which about 200 gm. will dissolve in 100 gm. of water. But if we compare calcite with feldspar, we find that calcite is many times as soluble. After all, even a slightly soluble material, acted on by water for a long enough time and in great enough volume, will be completely dissolved. This is the case in nature.

There is, however, an important factor that tends to make calcite even more subject to attack by solution than the figures just given would indicate. When there is any carbon dioxide in the water, then the solubility of the calcite goes up by leaps and bounds. That is because the calcite is converted into *calcium bicarbonate,* $Ca(HCO_{3)})^2$, as expressed in the reaction:

$$CaCO_3 + H_2O + CO_2 \rightarrow Ca(HCO_3)_2$$

Calcium bicarbonate has a solubility of about 0.04 gm. per 100 gm. of water, which makes it about forty times as soluble as calcite itself. It takes no expert to see where the carbon dioxide comes from. As the rain falls through the atmosphere, it dissolves some of that gas from the air; and as it percolates through the rocks, any calcite in its path soon goes into solution.

Suppose rain falls on an area in which limestone lies just beneath a thin layer of soil. Near the surface, where the loose soil affords an easy path for the water, there will be a percolation of carbon-dioxide-charged water coming in contact with the limestone. As the water runs down the cracks in the rock, some of the limestone is dissolved, the cracks are widened, and there is an even easier path for later rain water.

The limestone also varies in permeability here and there, as all rocks do, so that in the course of time the percolating waters will dissolve out a regular series of passages. When they do, caves and caverns result. Surprisingly enough, these scenic underground features, which annually attract so many tourists, are due chiefly to the unobtrusive work of ground water operating on soluble rocks.

As a general thing, caves are in those zones where the circulation of the ground water is most pronounced. That is, the lower levels of a series of caverns tend to be adjusted to the level of the major stream in the area. If the caverns become large, and are dissolved out near enough to the surface, their roofs may lack support and consequently fall in.

The steep-walled circular or oval openings that result are called sinkholes. They are relatively common in some cave areas, as we should expect. Sinkholes may also be formed by the enlarging of crevices leading down into the limestone; but whatever their origin, they act as funnels to lead water down into the rock layers, and thus help to carry on the work of solution.

In regions having numerous sinkholes most of the rain that falls on the surface may run down into the sinks, with the result that surface streams may be largely or entirely absent.

Almost everyone who has visited a cave or who has seen pictures of them seems to be more impressed by the curious for, mations in them than he is by the cave itself.

From the roof of the cavern may depend long slender "icicles" of stone; and upward from the floor spring similar, but generally more squatty, grotesque deposits.

These formations are only indirectly due to the solvent action of ground water; they are the direct results of its depositional activity. The "stonecicles" that hang from the ceiling are called stalactites, and those that rise from the floor are stalagmites.

In some eases they join; others are sheetlike instead of cylindrical: but one and all are due to a redeposition of calcite dissolved elsewhere by the ground water. See Plate 10 for photographs of these striking cave features.

Most caves have entrances, and so there usually is air circulating through them. Thus there is a chance for evaporation. If rain water, circulating downward through the rocks above a cavern, dissolves some calcite along its journey, it may drip from the roof of the cave. As it reaches the air it evaporates in part; and when it does, calcite is deposited. Thus by a slow process stalactites are formed and elongated. Meanwhile the water that drips to the floor may evaporate there, and thus build up a stalagmite.

As usual, this simple picture is complicated in nature, and we find that relative concentrations of carbon dioxide, pressures, and especially temperature changes enter into the interplay of factors that decide whether the ground water is to be in a dissolving or depositing mood. Again we must forego technicalities, but we can explain that if the environment entered by the water is one of relatively high carbon dioxide concentration, solution will be the tendency, whereas a lower concentration of carbon dioxide will cause deposition. There really is a nice balance here in any given cast and it affords one more example of the approaches toward equilibrium that Nature seems so anxious to achieve.

Those among our readers who like to study chemical equations will be interested to note that the following reaction really expresses the relationship between ground-water solution and deposition: For example, if conditions tend toward an increase of CO_2, the lower reaction is driven in an upward direction, and this drives the upper reaction toward the right.

Ground-water work is not confined solely to the solution and deposition of calcite, for it plays an important rôle in many geological phenomena which can scarce be mentioned here. For example, ground-water deposition of minerals about some nucleus may form *concretions* which range in size from microscopic objects to symmetrical structures a dozen feet across.

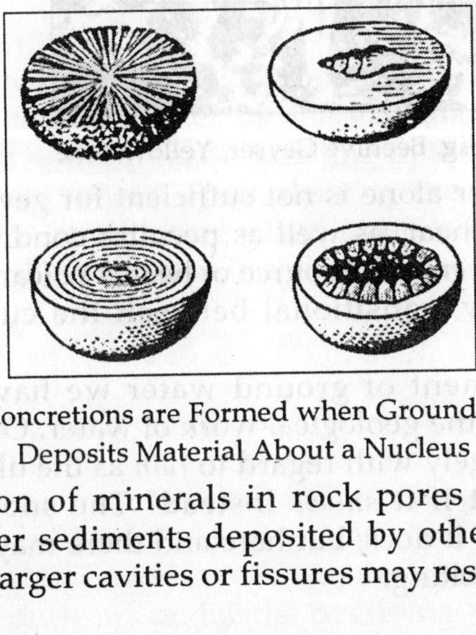

Fig. Concretions are Formed when Ground Water Deposits Material About a Nucleus

The deposition of minerals in rock pores is responsible for cementing together sediments deposited by other agents. Ground-water deposits in larger cavities or fissures may result in the formation of *mineral veins*.

Furthermore, deep beneath the zones of ground-water circulation, there may Still be a diffusion of dissolved materials; and when such diffusing solutions meet, chemical reactions may take place. Thus are formed some types of ore deposits. In fact, very commonly ground water acts to concentrate otherwise thinly distributed ores, and so contributes definitely to human economy.

Geysers, those openings in the earth from which steaming hot water is ejected at intervals, are more spectacular evidences of ground-water activity. Yellow stone Park is the region par excel. lence for these phenomena, and annually thousands of tourists be come acquainted with this relatively rare aspect of ground-water work.

Fig. Beehive Geyser, Yellowstone

But ground water alone is not sufficient for geyser action: there must be a source of heat, as well as peculiar conditions among the fissures in the rocks. Since the source of heat is volcanic, the subject of geysers is obviously transitional between the current topic and volcanism.

With our treatment of ground water we have finished two important aspects of the geological work of water. Our consideration thus far has been largely with regard to *rain* as the ultimate source of the water. But what if it *snows* instead? The snow may melt or evaporate, as most of it does; but here and there may be an excess of accumulation over melting.

RIVERS OF ICE

Nature is infinite in her moods. Consider water, for example. Under moderate temperatures it tumbles along the surface, constantly adjusting itself to irregularities in its path, seeking ever the lowest level it can find. Raise the temperature somewhat and the water vaporizes, disappearing into a tenuous gas which finds itself at the mercy of even the most capricious winds.

But now reverse the conditions, allow winter to hold the water in its frozen grip, and the turbulent spirit of liquid or gas is stilled. But is all movement necessarily lost? By no means. Almost everyone has read about the movement of those great rivers of ice called glaciers. Yes, given suitable conditions, even the apparently solid ice may flow; and on this curious planet of ours it is not so unusual to find just such favorable conditions for glacial movement.

When precipitation takes the form of snow instead of rain, the blanket of snow does not run off down the slope, but accumulates until it evaporates or melts; in the latter case the *meltwater* at once runs off. Now under certain favorable situations, such as high latitudes and high altitudes, the heat of the summer sun is not great enough to melt or evaporate all the snow of the preceding winter. Hence there is an accumulation of snow from year to year, which finally assumes considerable proportions.

Large snow fields are built up in this manner. Their lower limits form the permanent snow line of the mountains. Near the Equator this snow line occurs only on the tops of the loftiest mountains; in the middle latitudes it is found at much lower elevations; and in frigid zones it may descend to sea level. Thus, from Equator to poles there is a descending surface above which snow may remain all the year around.

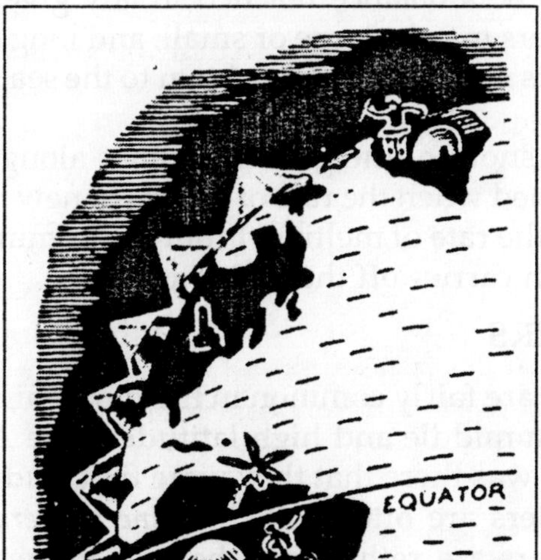

Fig. The Snow Line Descends from the Equator Toward the Poles

Although all the snow does not melt, some of it does, and part of the meltwater, percolating down into the snowbank, recongeals to form granular particles of ice. Similarly, the pressure of the overlying snow aids in converting the deeper layers into ice granules, which pass downward into porous ice; and toward the base, where the pressure is greatest, the ice may be densely compacted.

The pressures developed by the overlying snow and ice tend to seek relief, and this relief is found by the lowermost marginal parts of the ice mass moving down the slope of the mountain. Thus are glaciers born, and where movement begins is commonly regarded as the beginning of the glacier.

Glaciers do not move with a velocity anything like that of water. The ice moves slowly and ponderously: it is a matter of feet per year rather than feet per second as in streams.

The movement of glacial ice, indeed, is no simple matter, as we shall shortly see; at present we are concerned mainly with the fact that it does move downslope.

As the glacier moves downward, it tends to follow any irregularities along the surface. Commonly, glaciers follow pre-existing stream valleys down the mountain side.

These we call valley glaciers. The distance downslope that the glacier can extend depends in part on the increase of temperature in lower altitudes and in part on the rate at which snow is accumulating in the snow field above.

There tends to be a balance reached in any given situation; and consequently glaciers may be large or small, and long or short. In high latitudes the glaciers may persist right down to the sea, and there break up into *icebergs*.

Elsewhere the snout of the glacier reclines along the valley, and equilibrium is reached when the rate of arrival of new ice to the margin is just balanced by the rate of melting. From the terminus of the glacier issues a river which carries off the meltwater.

VALLEY GLACIERS

Valley glaciers are fairly common in the mountainous areas of the world, especially in middle and high latitudes. The Alps are famous for their glaciers; it was there that they were first studied, and for that reason valley glaciers are often called *Alpine glaciers*. In the United States valley glaciers are rather rare, being confined largely to the higher peaks of the Cascade and Wind River Mountains. Farther north, in British Columbia and Alaska, there are thousands of them, many of which persist down to sea level.

The similarities of valley glaciers to rivers may be seen by considering the snow fields above, which constitute their catchment basins; their confinement to valleys; and their movement downslope.

In addition, valley glaciers may have tributaries, so that two or more glaciers may merge to follow along a single valley from their point of juncture. In contrast to these similarities, of course, there are striking differences.

In addition to valley glaciers there are much larger masses of glacial ice called continental glaciers or *icecaps*, which cover all elevations of the region. Antarctica is covered with such an icecap, as is the greater part of Greenland.

In fact, an area about twice the size of the United States, or nearly 6,000,000 sq. mi. of the earth's surface, is today covered by such ice. These icecaps are much thicker at the centre than at the edges, so that there is a surface gradient from the centre outward.

This surface gradient causes a slow movement in such ice masses also, and the position of the outermost edge again represents an equilibrium between the rate of movement and the rate of melting, or the rate of breaking into bergs at the seaward margin of the ice field. Present-day icecaps are largely confined to the high latitudes, but in the geological past they covered much larger areas of the globe.

Glaciers, like rivers and wind, wear away the surface over which they move. The débris they pick up is largely carried to the terminus of the glacier, and there some of it is delivered to the streams of meltwater that flow on down the valley.

Much of the rest of it is literally dumped near the margin of the ice as a heterogeneous pile of débris. In many respects, however, the carrying of material by glacial ice strikingly differs from transportation by turbulent agents. This difference depends largely on the motion and physical properties of glacial ice as compared to those of other agents.

In the first place, glaciers move ponderously downslope at a slow and snail-like pace. Furthermore, glaciers are composed of solid ice granules, and thus are not liquid. As a matter of fact, as far as direct observation goes, ice behaves as a solid; its outer portions yield suddenly and discontinuously, like any other brittle substance. Great slicing or shearing movements transect the ice at an angle, so that masses of ice are literally shoved over the lower portions, as the glacier yields to the forces operating on it.

In Detail the structure of the ice often shows thin laminae arranged essentially parallel to the bed of the glacier. Relative movement takes place along these laminae, which may be zones of weakness in the ice. There is a suggestion that these laminae may also reflect a more continuous plastic yielding within the body of the ice, similar to laminar motion.

The net effect of all factors involved in glacial movement is a complex yielding to gravitative forces; and whatever else may be involved, turbulence, in the sense that we have used it, plays no part in the story. Furthermore, there is no sorting effect in glaciers.

In essence the ice simply freezes over, and carries with it, every movable thing, including rock masses literally torn from the glacier bed. Obviously, the shapes, sizes, or densities of the débris are not the determining factors in such transportation.

The absence of selectivity means that there is no traction or suspension load in the ice, in the sense that we earlier discussed. Ice, in fact, carries a heterogeneous load of débris of all sizes and shapes, intermingled haphazardly. The load ranges from immense blocks of rock down to the finest clay particles; and although much of it is concentrated at the base, yet in valley glaciers part of the load results from blocks falling on the surface.

Similarly, the slicing or shearing movements within the ice carry some of the basal portions of the load up into the main body of the glacier, and even to its surface. In contrast to the heavy loading in some zones, there may be layers of ice almost entirely free from débris, but this is rarely or never true of the basal portions of the ice.

If the frictional resistance of the basal load against the glacier bed becomes very great, the ice may be unable to move the load along. Then that basal portion of the ice becomes stagnant; and the overlying ice, less heavily loaded, may slide over it, using the basal layer as its new bed. Such complications naturally render an exact evaluation of all the factors involved in ice movement a bit difficult.

Since glacial ice carries its load in a frozen grip, the fragments at the base of the ice are held rigidly; and as they are shoved along the glacier bed, they literally gouge out great scratches and grooves in the underlying rock. It is easy to see that a really good-sized glacier may account for a tremendous amount of erosion under these conditions.

A glacier is thus like a huge plane scraping off the bedrock over which it moves. Even large rock hills are profoundly modified or entirely worn away by the overriding ice. If a part of the hill remains, the side which faced the ice movement is, in most cases, decidedly rounded off.

It might be predicted that the erosive power of ice should give rise to typical land forms, and detailed investigations have shown that it does. In the first place, valley glaciers scour out the valleys they occupy, and change the cross sections of water-carved valleys considerably.

Instead of being confined in their erosive action to the bottom of the valley, as water is, the glacier may occupy large parts of the valley. Thus the ice has an opportunity of cutting the bottoms and sides simultaneously, and in this manner it produces a typical valley form of its own. We say that glaciated valleys are like broad round U's in cross section.

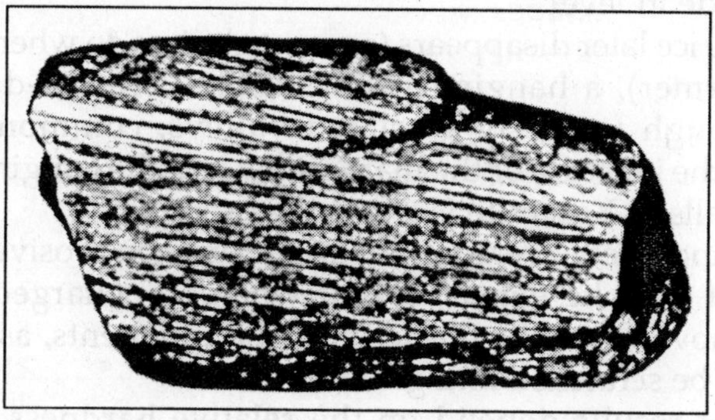

Fig. Glaciated Pebbles are Marked by Parallel Scratches

At the heads of valleys occupied by glaciers are huge amphitheaters called cirques, They are due to erosion, occasioned apparently by a plucking action at the head of the glacier. Alternate thawing and freezing, combined with an accompanying movement of compacted snow and ice, result in the excavation of a notch in the mountain side. As this increases in size, it develops steep walls and ultimately becomes a conspicuous feature in the landscape

Fig. Bridal Veil Falls Plunges from a Hanging Valley

Since some glaciers are larger than others, cirques vary

considerably in size. More than that, glaciated valleys also differ in the degree of glacial scour.

Thus it happens that where two glaciers merge, or where a glaciated valley passes an unglaciated tributary, the larger valley may be carved more deeply into the rock, so that the bottoms of the valleys do not coincide in level.

When the ice later disappears (as even glaciers do when the climate becomes warmer), a hanging valley results. Present-day streams, flowing through formerly glaciated valleys, commonly develop waterfalls at the juncture of the main valley and the hanging tributary. Bridal Veil Falls in Yosemite is a striking example.

Plate 13 includes several of the more common erosive features of glacial ice. It is to be expected that when ice, charged with rock fragments, moves over the glacier bed, the fragments, as well as the floor, should be scratched and ground.

The exact results depend on the relative hardness of the rock masses; but in general we do find within ice-transported débris a large number of pebbles with somewhat flat or soled sides, bearing sets of parallel scratches on them.

These *soled pebbles* are evidence that the ice held them rigidly while it moved along, otherwise the sides would not be flat, nor would the scratches be parallel. In some cases there are two or more intersecting sets of parallel scratches on the same face, indicating that as the pebble was moved along it struck one or more snags and was slightly rotated.

We may see now that glacial ice is individualistic in its movement, in its mode of transporting rock débris, and in the land forms that result from its erosive action. In fact, each geological agent has associated with it certain land forms and deposits; and from the geological point of view, the important thing is to recognize and understand how these features arise.

Thus far we have considered four of the five geological agents, largely in terms of their transporting work and the erosive land forms that accompany the removal of rock débris. We have yet to consider the opposite phase of the subject, namely, the *depositional* features of these geological agents. In the long run, erosion is balanced by deposition in nature; and it will be convenient for us now to turn our attention to this opposite aspect of our subject.

END OF THE LINE

What goes up must come down" is a homely old saying, but it applies quite well to the transportation of rock débris. The material that is picked up at any point by wind, water, or ice must be deposited somewhere else. This is inevitable because the energy of the transporting agent is limited, and eventually this energy is dissipated in the very act of transporting or by the many sources of friction that the agent encounters.

The transporting-capacity of a stream depends in large part on its velocity. As long as the velocity is increasing, the load may also be increased. If the velocity remains constant, the stream may continue to pick up material until it is utilizing its maximum carrying-capacity, when it is said to be *loaded*. If the velocity decreases after this loaded state is reached, some of the débris must be dropped. Consequently deposition also depends on the velocity, but in quite an opposite sense from transportation.

In the upper reaches of a stream valley the gradient is steeper than it is farther along its course. Now, in a perfectly frictionless fluid the velocity would continue to increase as long as any slope remained, because there always would be some component of the force of gravity operating on the fluid. In nature, however, the sources of friction are legion, and a stream's velocity tends to decrease as the slope becomes more gentle. There is an inner friction within the water itself, turbulence consumes some energy, and there is the friction of the moving mass of water against the banks and bottom of the stream.

The tendency toward decreased velocity due to energy losses through friction is in part counteracted by an increase in the volume or *discharge* of the stream as it flows along its course. An increase in discharge tends toward increased velocity.

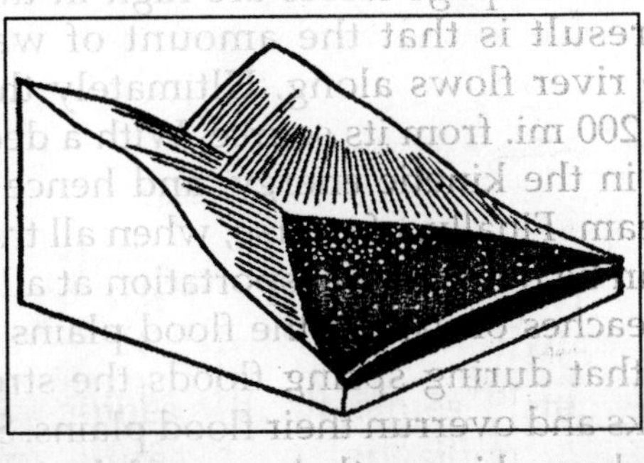

Fig. Block Diagram of an Alluvial Fan at the Mouth of a Mountain Valley

It may also act in the opposite sense, however, because a decrease in the amount of water will decrease the velocity of the stream. Evaporation is constantly taking place along rivers, and in some cases they may lose part of their water because of seepage through the bed. There is thus a balance of forces in any given case which tends to determine the conditions of transportation or deposition that will apply. Now, although the velocity determines whether transportation or deposition will take place, the velocity itself depends largely on changes in the gradient, changes in the volume or discharge of the water, and in the friction it encounters from its channel.

We should accordingly expect sites of deposition to be those very places where the stream gradient decreases appreciably, where decreases in volume take place, or where the frictional effect increases. The best examples of gradient decrease are to be found at the bases of mountains. A mountain stream, tumb down a rather steep gradient with a high velocity, carries a heavy load of débris, as it actively erodes its channel. At the base of the mountain the land surface flattens out rather abruptly into a plain.

This relatively abrupt change in the gradient results in quite as sudden a decrease in the stream velocity, and hence there is deposition at this critical point. In fact, large aprons of deposited material, which spread out in fanlike forms, are found at the foot of many steep slopes where streams issue from the mountains. These deposits are called alluvial fans and are due to conditions precisely like those outlined above.

Another example of deposition, involving a change in volume (and hence in discharge) of the stream, may be mentioned. Some streams arise in regions of appreciable rainfall, and during their course flow across lands that are more arid. Such a stream is the Humboldt River, which rises in the northeastern part of Nevada and flows over the arid plains of the Great Basin.

Evaporation and seepage losses are high in the arid part of its course, and the result is that the amount of water continually diminishes as the river flows along. Ultimately the river dries up completely, about 200 mi. from its source. With a decrease of volume comes a decrease in the kinetic energy, and hence in the carrying-capacity of the stream. Finally, of course, when all the water has been dissipated there can be no more transportation at all.

In the lower reaches of valleys the flood plains are usually very wide and flat, so that during spring floods the streams commonly overflow their banks and overrun their flood plains. Since the velocity of the stream in its channel is greatly increased during such highwater stages, the floodwaters may be carrying a considerable load of débris. Now the moment the flooding begins, the sheet of water

That spreads sideward out from the main channel meets with considerable friction along the bottom, and thus is greatly retarded. This is due to the large area of contact which such a shallow sheet of water has with the flood plain. What happens is simply explained: As soon as the velocity is checked, some of the load is dropped; and since the greatest checking of the velocity takes place right at the main channel edge, it is there that the bulk of sediment accumulates.

Fig. Block Diagram of a Natural Levee

Thus are built up natural levees, or low, broad ridges of débris along the channels of these old streams. When the volume of water in the channel again subsides, these levees stand above the new level of the stream and serve to confine the water within their bounds. People living along many old valleys build up these natural levees to greater heights, in an attempt to prevent the floodwaters from flowing over onto the flood plains. The lower reaches of the Mississippi Valley afford good examples of such levees, and everyone is familiar with the havoc raised when destructive floods do tear away or overflow these ridges of earth.

The building of natural levees results from a velocity decrease due to frictional forces. Similar frictional situations also occur at the mouths of streams, where rivers empty themselves into lakes or seas. What happens is that the velocity of the river is checked by the inertia of the relatively quiet waters into which it flows. Thus there is a dissipation of energy due to friction. The Mississippi River will again serve here as a good example.

The river empties into the Gulf of Mexico; and as it does so, the velocity of the water is checked to such an extent that most of its load is dropped. In the absence of strong currents along the shore at that point, the deposited material accumulates near the mouth of the river. In the course of time a delta has been formed, which grew out into the bay as a fanlike deposit spread out widely from the river's mouth.

Such deposits usually assume the form of a rough triangle, which is the shape of the Greek letter delta (Ä); and from this symbol they derive their name. In the formation of deltas an equilibrium is maintained between the deposition of the load that is continually arriving and the material already deposited.

As the delta grows in length, its upper surface, being essentially at sea level, must be adjusted to allow the débris to be carried out to its edge. Thus it is necessary that a slope develop in the direction of the river's flow. This slope is formed by the deposition of débris toward the landward edge of the delta, and in this manner the upstream portions of the deposit gradually emerge above sea level. Plate 14 includes some of the more common features formed by stream deposition.

When the differences between traction and suspension loads are taken into consideration, we find further interesting details arising concerning sedimentary deposits. The suspension load is generally finer than the traction load, which we saw consisted of particles too large to be picked up and yet small enough to be rolled along the bottom. In any given stream—or air current, as a matter of fact—there is usually a gradation between the two types of loads. The traction load, however, is the marginal portion, and it will tend to be dropped at the slightest decrease in velocity. This effect is brought out most strikingly in the case of wind. A current of air carrying dust in suspension and sand as a traction load may move over an area where vegetation tends to slow down the velocity of the wind, As soon as the velocity is decreased, the traction load is deposited, but the suspension load may be carried right over the obstacles and so escape deposition. The fact that the traction load of sand is stopped by obstacles means that the sand will be deposited in mounds or hills; and as soon as some sand is dropped, it acts as an additional barrier. In this manner sand dunes are built; and such deposits are definitely localized as mounds, hills, or ridges, depending on the supply of sand, the direction of the winds, the obstacles, and other factors. Sand dunes are fairly common land formes, and they are generally found near sources of loose sand. Hence we find them along coasts and also in arid regions where windwork has an opportunity to manifest itself. Dunes have typical forms which shed light on their mode of formation and their slow migration over the surface. For dunes do move; and cases are not unknown where even towns have been engulfed by the sand, only to appear again when the sand moves beyond.

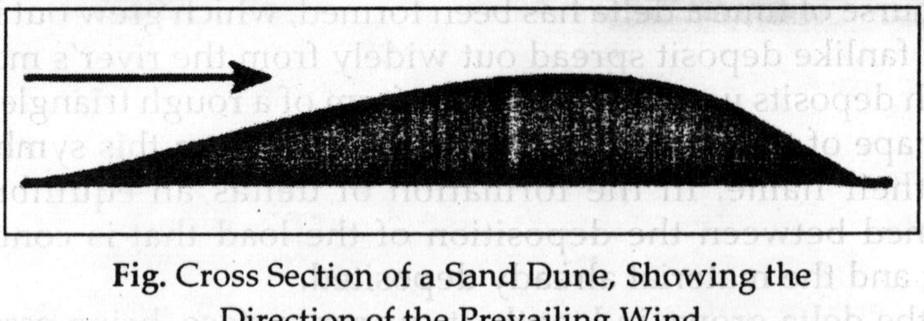

Fig. Cross Section of a Sand Dune, Showing the Direction of the Prevailing Wind

In profile the dunes present a gentle slope to the wind, and a steeper slope to the lee. The grains of sand are blown up the gentle slope and tumble over the lee edge, where they lie at the angle of rest of loose sand, about 30°. The continual movement of sand from the front to the rear slope of the dunes results in a slow movement of the dune as a whole. Many dunes are rather low features, but hills up to 200 ft. high are not uncommon.

On a much smaller scale are the tiny *ripples* that commonly form on dry sand; and if they are watched, it is seen that they move with the wind.

Here also is a migratzion of the sand over a gentle windward slope and down a steeper lee slope. The development of the ripples appears to be a response to conditions demanding minimum friction between the moving air and the semi-mobile grains of sand.

Between the ripples are tiny horizontal eddies, which allow the wind to blow over the crests and over these horizontal eddies with less friction than would be encountered over a smooth field of sand. Plate 15 illustrates these and other depositional features of the wind.

The suspension load of dust in the air is carried beyond the localized obstacles and remains in suspension as long as the wind has enough motion to keep the particles off the ground.

Fig. Loess Stands in Steep Walls During its Subsequent Erosion

When the wind dies down, as winds inevitably do, owing to energy dissipation, the suspension load also tends to settle, not locally in mounds, but over a wide area. The net effect is that the suspension load forms a mantle of fine material over the landscape, which may be fairly uniform in thickness over large regions. There is a common wind deposit called loess, with exactly these characteristics. It is formed from the suspension load of winds, and it is disseminated widely as a blanket which, unlike other deposits, clothes hills and valley slopes indiscriminately. Loess has the interesting characteristic that it is capable of standing in very steep walls when it is itself subjected to later erosion.

Such simple analyses as the foregoing indicate the importance of distinctions between traction and suspension loads, because in the interpretation of deposits they afford a means of determining the conditions that prevailed during the formation of ancient rocks on the earth. We must not gain the impression from our earlier discussion of traction and suspension loads that the materials deposited from the traction load are always quite spherical or that those from suspension loads are always flat.

During transport even the partitles in suspension may become more rounded from mutual impacts and other causes, and new loads are often picked up in the course of travel The net result is that the final deposit may have quite an assortment of various shapes and sizes.

There is an extremely important characteristic of deposits laid by turbulent agents, and that is the sorting of the deposits.

From the miscellaneous débris made available for transport by weathering and erosion, the turbulent transporting agents sort out, or select, particular sizes, shapes, or densities by virtue of the differences between the traction and suspension loads, as well as by the carrying-power of the agent itself.

When later these loads are dropped, the deposition is also somewhat selective, as we have already seen. The net result is that there is a fairly definite sequence of deposits sorted according to size and composition in any depositional environment. Thus arise the materials that we call *gravel, sand,* and *mud*.

It is important to see that only through sorting action is it possible to separate the miscellaneous débris of weathering into a series of sediments having restricted ranges of sizes, shapes, or densities. Further, the particular agent that does the sorting often leaves other marks of its action on the material, and these furnish us with additional clues regarding ancient sediments.

Another striking feature of deposits left by turbulent agents is bedding. We saw that decreases in velocity cause selective deposition, so that a layer of débris is laid down. Later a slight shift in conditions or a difference in the load may bring about the deposition of another layer. These layers of material may then be separated by a distinct juncture plane, known as a bedding plane.

The layers between these planes are called *beds*, or strata; and consolidated or cemented deposits showing such features are called stratified rocks. These rocks, composed, as they largely are, of fragments of weathered and broken deep-seated rocks, are part of the group of *fragmental rocks*.

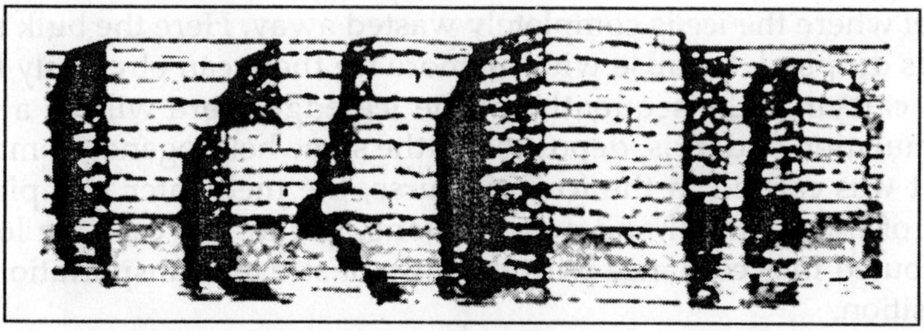

Fig. Horizontal Bedding and Vertical Joints in Sedimentary Rock

Stratification of the sort just mentioned is due to selective transportation and deposition. But all stratified rocks are not necessarily fragmental rocks.

Successive lava flows, poured out of volcanoes, may harden into layers of rock which also have a stratified appearance, although they lack the other characteristics of sediments. We should understand from this, however, that stratification is not confined to the sediments; indeed, not even all sediments are stratified.

In this connection the deposits left by glacial ice are in striking contrast to the bedded deposits of wind and water. The fact that ice is incapable of sorting out the load that it picks up means.

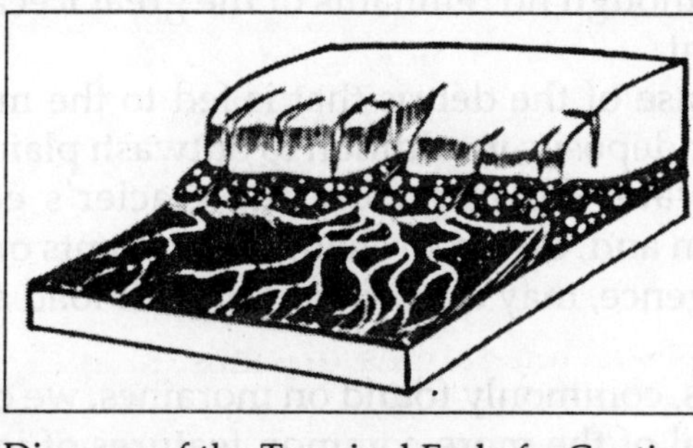

Fig. Block Diagram of the Terminal End of a Continental Glacier

The morainic ridge lies beneath the edge of the ice, and farther back under the ice is the till plain. The outwash plain, with its streams of meltwater, is in the foreground that the deposits left by the ice must necessarily reflect the same heterogeneity as the original débris.

This situation is, in fact, met in nature, because the most outstanding feature of ice deposits is, that they are unsorted. In the deposits we find a heterogeneous mingling of all sizes and shapes, just as it was left when the glacier melted and dumped its load at the point of melting. This heterogeneous sediment laid by glacial ice is called glacial till.

Suppose a glacier moves down slope with its load until it reaches a level where the ice is completely wasted away. Here the bulk of the load is dropped, because with no more ice there can obviously be no more carrying. Consequently, at the ice edge there will be a great accumulation of débris, deposited in the same heterogeneous manner that it was carried in the ice. Of course, the meltwater will pick up some of this material and carry it along, but in this case the load is distributed in accordance with the laws of stream transportation and deposition.

What we find then is a large ridge of unbedded and unsorted débris at the edge of the glacier, and stretching out from this ridge a series of bedded sands and gravels that have been carried away by the meltwater. Such ridges built at the edge of the ice are called moraines, from their wall-like aspects; and the gravels and sands beyond them are called outwash plains.

Not only is glacial till piled up in moraines along the ice edge, but it may also be deposited as a mantle beneath the ice behind the moraine. Such deposits are called till plains because they commonly occur as relatively smooth or gently rolling plains after the ice has disappeared. In fact, large areas of northern North America and Europe are covered with till plains, which are important evidences that these regions once were glaciated, although no remnants of the great icecaps now remain on either continent.

Even in the case of the débris that is fed to the meltwater by the ice, there are other deposits in addition to outwash plains. For example, streams of meltwater flowing from the glacier's edge may carry material with them and, owing to changed gradients or pressures near the point of emergence, may deposit part of their load as hills of gravel or sand.

These features, commonly found on moraines, we call kames. Plate 15 includes several of the more common features of ice deposition.

In fact, the work of glacial ice affords an interesting example of the great complexity of nature when examined in detail. The intermingling of ice-laid and water-laid deposits is so typical of glacial phenomena that we usually think of the associated outwash deposits as a part of glacial action itself. In a more advanced text we would call the outwash deposits *glaciofluvial* deposits, to indicate that both ice and water had a hand in their formation.

Just as there is a close relation between the geological agents and their deposits in nature, so there may be a close relationship between erosion and deposition itself.

Thus far we have kept transportation and deposition somewhat separate, but that has been largely to keep unity in our discussion. We shall combine the two, however, in turning to the next geological agent, and endeavor to show how both erosion and deposition may take place side by side in the development of certain land forms.

Chapter 9

Nature of Glacier Flow

Glaciers are slow-flowing, thick masses of ice. *Snowfields* are thinner and essentially motionless masses of permanent snow. Snowfields and glaciers are savings banks in the water economy of the earth. Each year, some of the water evaporated from the seas and lands falls as snow.

Most of the snow melts in summer, but in polar regions and at high altitudes part of it remains throughout the year in glaciers and snowfields. If withdrawals by melting and evaporation exceed deposits as snowfall, the balance in the bank decreases and the glacier recedes, but if precipitation exceeds withdrawals, the glacier grows and spreads outward.

THE SNOWLINE

The lowest altitude at which permanent snow rests is called the *snowline.* Obviously, it does not persist always in one position but is affected by variations in the annual snowfall, temperature, wind direction (which controls drifting), and topography (which controls both avalanching—snowsliding —and shading from the sun).

Mean annual temperature decreases at higher altitudes and at higher latitudes, hence the snowline is higher near the equator than in polar regions.

The amount of snowfall also affects its position. Thus, the snowline stands over 5,000 feet higher (8,000 to 9,000 feet) on the dry eastern side of the St. Elias Mountains along the Alaska-Yukon border than on the wet western side (2,500 to 3,000 feet), and it is lower in well-watered Norway than on the colder but dry Taimyr peninsula. In much of Siberia, northern Alaska, and Canada, the mean annual temperature is low enough, but the precipitation is too scanty to support permanent snowfields and glaciers. In such areas, the pore moisture in the soil and rock is frozen to considerable depths, forming a great sheet of *permanently frozen ground.*

SNOWFIELDS

Above the snowline, on all except the steepest and windiest slopes, lie permanent snowfields. Excavations show that the beautiful geometric patterns of new-fallen snowflakes do not persist at depth. Instead, the snowfield consists largely of small rounded granules of ice about the size of birdshot. This material, called *firn*, has been formed by compaction of the feathery snowflakes and by alternate thawing and refreezing of their edges.

This melting and refreezing is not due entirely to variations ill air temperature above the snowbank. Water expands 9 per cent when it freezes; hence, pressure tends to lower the melting point of ice. Thus, even at temperatures slightly below the normal freezing point (0° Centigrade or 32° Farenheit), the snowflakes deep in a snowbank may be so tightly packed by the weight of later snowfalls that a thin film of water forms at the points of contact between them.

This water film may then move to a nearby pore space, where the pressure is lower, and refreeze into ice, even though there has been no change of temperature.

Rains falling on the snowbank and meltwater formed on a warm day are absorbed as in a blotter by the granular firn. Perhaps the following night this water freezes, assisting in the compaction of the firn by excluding air spaces. Thus, deep in the snowfield the firn gradually changes to interlocking granules of solid ice. Crevasses in glaciers commonly show all steps in the transformation: Snowflakes '! Firn '! Solid Ice.

GLACIERS

If snow and firn merely continued to accumulate each winter, even the highest mountains in arctic regions would eventually be buried. Ice, however, is not a strong mineral, and its strength limits the height to which it can accumulate.

Before any great thickness is reached, the ice at the bottom begins to creep and flow outward under the overlying weight. Thus, from the snow and firn a *flowing ice mass,* or *glacier*, is formed.

GLACIER MOTION

Glaciers move so slowly that the motion cannot be seen, but the fact that the ice actually is moving can be readily proved by driving a straight row of stakes across the surface of the glacier. In a few days or months the straight line becomes a curve, bent downstream in the centre of the glacier where the speed is greatest.

The rate varies in different glaciers and also with the season of the year. In a few of the Alaska and Greenland glaciers, velocities of more than 150 feet per day have been recorded, but such rates are exceptionally high. The lower parts of some glaciers on the coast of Alaska appear to be practically stagnant.

Indirect proof of glacier flow is given by the rocks frozen into the glacier. Most large glaciers contain abundant boulders which must have been carried from far upstream by the moving ice, because no similar rock is found in the valley walls nearby. The bodies of climbers who crashed to their death in the treacherous crevasses of the Bossons glacier in the Alps were released 41 years later at the terminus of the glacier several miles below.

Nature of Glacier Flow

The upper surface of glaciers is riven by crevasses, and the ice appears to be hard and brittle. It shatters like glass under a sharp blow, just as does an ice cube from the refrigerator. It is difficult to imagine such a substance flowing. Actually, there is little or no evidence that the brittle, upper part of a glacier does flow.

The abundant crevasses prove that fracture is the chief form of yielding. Deep within the glacier, however, deformation is almost entirely by flowage. Glaciers are perfect examples of the effect of scale on strength . Ice in small masses is rigid and brittle, but in glacial masses it flows under its own weight. The brittle, crevassed surface ice is rafted along on the flowing ice below.

The mechanism of glacier flow is complex. Microscopic studies of ice obtained from glaciers show that some of the ice crystals have been bent, others have glided along the sheets of atoms parallel to the base of the hexagonal crystals, and still others have been granulated and sheared. Furthermore, the crystals tend to be arranged in parallel planes, as in foliated metamorphic rocks, suggesting that they recrystallized during or after movement.

Such recrystallization is demonstrated by the coarser crystals of glaciers as compared to those in snow and firn. Crystals in the firn of Alpine snowfields average less than ¼ inch in diameter, but at the glacial snout the grains average about an inch in diameter. Crystals 6 to 8 inches in diameter may be collected from the edge of the Malaspina glacier in Alaska. The large crystals must have grown by recrystallization of smaller ones.

The ice at the front of a glacier, or in deep glacial crevasses, is generally layered. The layering superficially resembles stratification. It lies roughly parallel with the floor of the glacier and curves upward along its walls. Measurements at the glacier front have shown that the ice in adjacent layers moves at different speeds, proving that the layers are separated by shear surfaces. This layering is one manifestation of shear flow—the layers are due to the development of many closely spaced surfaces of shear within the glacier along which the ice and contained rock detritus are dragged forward and streaked out by movement.

The shear-banding of glaciers closely resembles the foliation of many gneisses and schists. Clearly, ice under a relatively small load is no longer brittle, but reacts to differential pressure by flowage and recrystallization. We have, in glacier ice, an example of a metamorphic rock, completely transformed from the original snow and firn into a wholly different streaky crystalline mass by movement and recrystallization under pressure.

KINDS OF GLACIERS

The form of a glacier is largely controlled by the topography over which it flows. *Valley glaciers* are streams of ice that flow down the steep-walled valleys of mountain ranges. Fed by snowfields on slopes and in catchment basins above the snowline, the glacier may extend far below the limit of perpetual snow.

All glaciers end at the point where the rate of melting at the front equals the rate of ice replenishment by flowage. Valley glaciers are typical of lofty mountain ranges the world over. Most of those in the United States, except for the extensive glacier system on Mount Rainier, Washington, are short blunt ice streams only a few hundred feet thick. It is difficult to distinguish many of them from snowfields; indeed there are all gradations.

The Rocky Mountains, the Cascade Range, and the higher parts of the Sierra Nevada contain hundreds of small, rounded or irregular ice masses that lie in steep-walled clefts or else occupy amphitheater-like depressions opening out over steep slopes or cliffs. Such masses are called *cliff glaciers* or *hanging glaciers*.

At the foot of the St. Elias Mountains in Alaska several of these long systems of valley glaciers emerge from the mountain front and spread out on the plain below to form the Malaspina glacier, a lobate mass of ice covering 800 square miles. Such glaciers are called *piedmont glaciers*.

The largest glaciers are huge ice sheets, called *continental glaciers*. They are found today only in the polar regions. All of interior Greenland (about 637,000 square miles) is covered by ice, leaving only a narrow fringe of land along the coast. The Greenland glacier spreads outward in all directions from two high points in the interior.

Locally, the coast is bordered by lofty mountains through which the glacier spills, splitting up between the peaks into valley glaciers which then descend the coastal valleys to the sea.

Antarctica supports a much larger continental glacier, estimated to cover 5,000,000 square miles—an area larger than that of the United States and Mexico. The Antarctic glacier overrides the coast and projects into the "shelf ice" formed by the freezing of the sea. In places, the Antarctic glacier is held back by mountains through which the ice escapes in a series of large valley glaciers.

The famous Beardmore glacier, ascended by several of the early explorers in their quest for the South Pole, is one of these outlet rivers of ice from the high polar ice mass. The Beardmore glacier is 300 miles long, 12 miles wide, and extends far out into the shelf ice even after it reaches the sea. Small masses of radially spreading ice are found on Iceland, Spitzbergen, in parts of Scandinavia, and on the islands north of Canada.

They are called *ice caps* to distinguish them from the huge continental glaciers. Glaciers are much more powerful agents of erosion than rivers, but, because ice is restricted to high altitudes and to polar regions, running water moves far more rock debris each year than does ice.

ACQUIRING OF LOAD

Frost Weathering and Avalanching

Glaciers acquire rock debris in several ways. Valley glaciers commonly cover only a small part of the mountains in which they lie. On the gentle slopes between and above them lie extensive snowfields, but on steep windswept slopes even snow is unable to cling throughout the year. Hence, great expanses of craggy ridges, peaks, and cliffs rise above the glaciers.

These bare rock surfaces are strongly shattered by frost action. During the daytime, meltwater from snowbanks seeps into the rock crevices, and then, perhaps that same night, freezes and breaks blocks loose from their parent ledges. Small grains, loose chips, and even great blocks of rock thus freed tumble down the steep slopes and accumulate in talus piles on the edge of the glacier. Much debris is also carried down by avalanches (snowslides) during the spring thaw. Freshets from melting snow or summer rains also wash much debris onto valley glaciers.

Gradually, by being buried under new accumulations of snow or avalanche deposits, and by falling into crevasses opened by the motion of the glacier, much of this rock debris becomes incorporated within the ice, especially along its edges. The edge of a valley glacier may contain so much debris that it shows up in photographs as a conspicuous dark streak.

These stripes of dirty ice and loose rock along the edges of the glacier are called *lateral moraines*. Where two valley glaciers unite, the adjacent lateral moraines merge and form a *medial moraine*. If tributary glaciers are numerous, many medial moraines may streak the surface of the main glaciers. The moraines are not just surface features, they are wedgelike masses that extend deep into the glacier.

Meltwater Shattering

The exposed crags above a glacier are subjected to strong frost action but a glacier blankets its floor from extreme temperature changes. However, the rock floor of a thin glacier may be exposed to meltwater at the bottom of crevasses, and here a change in temperature will cause some frost shattering. At the head of most valley glaciers is a deep arcuate crevasse, or series of closely spaced crevasses, called the *bergschrund*.

The bergschrund gapes open in summer, but is filled or bridged over by snow in winter. It is inferred that the bergscbrund is formed by downstream movement of the plastic ice in the lower part of the glacier which breaks the brittle ice above and pulls it away from the rock wall.

Adventurous observers have descended into bergschrunds on ropes and found the lower parts to have a rocky wall on the upstream side and an ice wall on the other. The rock wall is shattered and riven by joints, some of which gape open and disclose slight displacements of the angular blocks of rock that they outline. Some of the blocks are free but have moved only slightly; others lean out against the ice in precarious unbalance; and the rest, entirely free from their parent ledges, are now incorporated in the glacier.

During the day, water from the melting snow pours over the face of the bergscbrund, filling the joints. The almost nightly frost loosens new blocks. Obviously, here is a zone of active glacial erosion, sapping the headwall by the plucking of frost-rifted blocks, steepening the cliffs behind, and gradually extending the glacial channel headward. An open crevasse, however, cannot extend to depths of more than about 200 feet.

Experimental tests on ice under pressure show that at loads corresponding to greater depths ice will flow plastically and slowly close the deeper openings. Nevertheless, certain typical features of glacier heads indicate that shattering somehow does go on to greater depths.

One factor is that the meltwater itself opens channels in the ice at the bottom of the bergscbrund. On hot summer days much meltwater cascades into the bergschrund. A "chinook" wind or a warm summer rain may greatly increase the volume. Despite this great volume of meltwater, the bergschrund practically never fills with water to the point where streams emerge and flow over the ice.

Apparently, the inpouring water melts its way down between the rock face and the glacier far below the bottom of the bergschrund, possibly, indeed, to the very base of the glacier. Actually, local melting of the deep ice should be expected from physical principles; the compressed ice at depth melts at a slightly lower temperature than the ice and snow at the surface.

Meltwater pouring into the crevasse also carries heat downward to melt more ice deep within the glacier. Plastic flow of the ice tends to close the meltwater tubes, but as long as heat is being continuously transferred downward, melting of the walls may offset the constriction by flowage. Furthermore, water in the tubes may be under high hydrostatic pressure. and the pressure opposes plastic closing.

Thus, we have a mechanism by which the rock wall at the head of a glacier may be wetted periodically to great depths. But, just as periodically, freezing cuts off the supply of water and the accumulated meltwater within the glacier and in the cracks in the rock wall may ultimately freeze. As a result, the rock wall is shattered and disintegrated to depths much greater than the bottoms of the crevasses. Frozen meltwater also adds volume to the glacier. Such a glacier, now frozen tight to the headwall, moves slowly forward, drawing out the shattered blocks of rock with it, and tending to open new cracks and crevasses to be filled by a further flow of meltwater.

It must be remembered, however, that meltwater in the deeper part of the glacier is protected from day and night changes in temperature. Deep within a glacier the temperature remains almost constant at the freezing point, and ice and water are in physical equilibrium with one another. Change from liquid to solid may occur here chiefly in response to variations in pressure.

A crack opened by the pull of the down sliding glacier will almost immediately fill with ice—the lowered pressure within the crack causes any moisture gaining access to it to freeze. But, in freezing, a gram of water gives up 80 calories of heat, and, if this heat is not dissipated, freezing will stop. For these reasons, meltwater shattering must be less effective at great depths than on the wall of the bergschrund.

Cirques

Downstream flowage of the glacier removes the shattered debris and exposes new rock surfaces to attack. Where valley glaciers have melted away completely, exposing their headwalls, the glaciated valley generally ends in a semicircle of high cliffs that bound an arcuate rock basin generally occupied by a beautiful mountain lake. Such cliffed valley heads are called *cirques* . The curving cliffs at the head of the cirque are shattered clear to their bases and they generally meet the smoothed and polished floor of the valley at a sharp angle. The jagged and shattered cliff face—the product of frost action and meltwater shattering-contrasts strikingly with the smooth valley floor—the effect of rasping by glacial ice.

THE GLACIAL RASP

The head of a valley glacier is a zone of rapid accumulation of snow and ice heavily charged with rock debris. If accumulation has been rapid enough to produce a steep downstream gradient of the top of the glacier, outward movement in a thin ice mass may occur along curving shear planes similar to the movement in a landslide. In a thicker and longer glacier, extrusion flow of plastic material may be less concentrated along shears, but it will follow similar paths.

As the ice moves across the bedrock floor, rock debris frozen in its base is dragged across the glacier bed, abrading the bedrock. The glacier thus becomes an extremely effective cutting and polishing tool; it acts like a gigantic rasp, gouging into and scraping off irregularities on its bed, and smoothing and grooving the rock beneath it.

Debris in Continental Glaciers

Practically the entire landmass occupied by a continental glacier is covered with ice; thus frost shattering and talus accumulation can take place only around relatively rare islands of rock that project through it. Also, little debris can be won by meltwater shattering of the rock floor at the bottom of crevasses. Yet, the ice at the margins of the Greenland glacier and other ice caps is heavily charged with rock debris, just as in valley glaciers. Some of this debris is the eroded soil that formerly covered the land over which the glacier flowed. In many localities, where continental glaciers have melted back along their edges, a striking feature of the rock floor formerly covered by ice is the almost total absence of residual soils and of preglacial unconsolidated mantle such as stream gravels and floodplain deposits. The soil and loose rock have been frozen into the lower part of the glacier and dragged off bodily by the moving ice. As in large valley glaciers, rock floors under continental glaciers are smoothed and polished. On these polished floors are numerous scratches and even deep grooves. Obviously, the glacier has gouged and abraded material from its bed.

Rate of Erosion by Glaciers

The rapidity with which a glacier abrades its bed depends on four factors: the resistance to abrasion and plucking of the particular bedrock, the abundance of cutting tools (rock fragments) frozen into the base of the ice, the speed of flow, and the weight (thickness) of the ice. Thus, thick continental glaciers flowing over weak or shattered rock are powerful agents of erosion. Similarly, on the floor of a cirque, glacial abrasion is generally great because here the ice is likely to be thickest and it is abundantly supplied with cutting tools.

DEBRIS RELEASED BY GLACIERS

Till. That glaciers are effective eroding agents is well shown by the vast amount of debris released at the glacier front by melting. Piled in hummocky ridges (*moraines*) along the front of the ice are heaps of boulders, sand, and silt, heterogeneously mixed without appreciable sorting or stratification.

Such unsorted debris, deposited directly by the ice, is called *till*. The proportion of boulders to fine material in till ranges widely. Some till is largely coarse boulders, but that from thin ice caps eroding shale and limestone may be chiefly clay and silt, with only scattered boulders.

Rock fragments found in till are not shaped like those in stream or marine deposits. Stream and beach pebbles are commonly rounded, but most of the fragments in till are subrounded or sharply angular. Some have been cracked and broken by the crushing weight of the overriding glacier; many, especially in valley glaciers, are joint blocks loosened by frost action and little modified during transport.

Some fragments show one or more nearly flat, grooved, and polished surfaces. Obviously, these flat surfaces were faceted and smoothed as the boulder was dragged against the bedrock floor, or against other rock debris within the ice.

Rock Flour

The effectiveness of the glacial rasp is especially obvious from the large amount of fine material (silt and fine sand) released at the glacier front. Turbulent streams of milky water generally cascade from one or more tunnels in the front of a glacier. If we collect a beaker of the roily water and let it stand, most of the suspended material promptly settles to form a thick layer of silt and fine sand on the bottom of the beaker. A very small amount of fine clay particles may remain suspended for hours or days. The silt, examined with a microscope, differs strikingly from the mud brought down by freshets in an unglaciated region.

The minerals composing the silt of the glacial streams are not the ordinary soil minerals characteristic of chemical weathering. Instead, small sparkling cleavage fragments of unweathered feldspar and other undecomposed minerals are abundant, and there is an almost complete absence of the yellow iron stain, the black and gray colors from humus, and the slimy clays so characteristic of chemical weathering. Clearly, most of this material is not an ancient soil removed by the glacier, but is *rock flour* made by the grinding and crushing of rock fragments against one another and against the bedrock.

Glaciers release an amazing quantity of rock flour. Glacial streams are almost invariably overloaded. They quickly deposit the coarsest material (rubble and coarse sand), building up a sloping alluvial plain over which they spread in braided courses. Most of the rock flour, however, is carried far beyond the front of the glacier and builds floodplain deposits farther downstream, or is spread into lakes as extensive delta and bottom deposits. Such rock flour forms great *terraces of white silt* along the rivers of British Columbia and Alaska.

The powdered rock is also widely scattered by the winds. Extensive unstratified deposits of *loess* (a wind-deposited soil composed chiefly of undecomposed silt particles) are found just beyond the limit of glaciation in many areas. Loess characteristically, contains abundant small tubelike holes where grass roots have grown.

FORMS OF GLACIAL DEPOSITS

The debris dumped by glaciers and by the streams flowing from them is called *glacial drift*. The drift assumes characteristic and easily recognized topographic forms.

Moraines

Unstratified glacial drift (*till*) is deposited directly by the ice and occurs chiefly in the form of moraines. The term *moraine* is used both for material deposited by the ice and also for the material upon or within the moving ice. The largest and best developed moraines are generally at the front of the glacier. If the ice front is in equilibrium, so that the rate of flow equals the rate of melting, it will remain essentially fixed in position for a long time. Here, the debris carried by the ice is released and accumulates in great hummocky morainal ridges. Because the ice front fluctuates with mild climatic changes, however, more than one morainal ridge is likely to form. The moraine that marks the farthest advance of the glacier is called the *terminal moraine*. Moraines that mark stages of halt as the glacier recedes are called *recessional moraines*.

Most terminal and recessional moraines of continental glaciers are broadly lobate in plan and can be followed, with minor interruptions, for miles. The moraine at the front of a valley glacier is generally a high crescentic ridge curving around the front of the valley glacier and extending up its sides as lateral moraines. Small recessional moraines of valley glaciers may be almost buried in outwash debris.

During the advance and withdrawal of a continental glacier some till is deposited beneath the ice and overriden by it. Till may also be scattered over a glaciated area by melting out of the englacial (ice-enclosed) debris as the ice recedes.

This irregularly scattered till, called *ground moraine*, is not concentrated into definite ridges like the terminal moraine. Some is found packed into hollows such as preglacial stream canyons; some is plastered around and upon low hills of the bedrock floor. Actually, the ground moraine is the most widespread deposit of a continental ice sheet. In most places it is spotty and only a few feet thick, but near the margin of the ice sheet it may cover hundreds of square miles in an almost continuous veneer.

In some areas, the ground moraine has been moulded into clusters of closely spaced hills, each shaped like the bowl of an inverted spoon, with the steeper slope facing the source of the ice. These streamlined hills are called *drumlins*. They vary in size, but are commonly about a mile long, ¼ mile wide, and 50 to 150 feet high. Drumlins occur in clusters of scores or hundreds. Each drumlin in the cluster has its long axis roughly parallel to the direction of flow of the former glacier.

Although we cannot see drumlins forming in modern glaciers, they are doubtless masses of subglacial debris overriden and moulded into streamlined forms by an actively flowing ice sheet. Excavations show they are composed largely of till, commonly a sticky variety containing much clay. Some drumlins have cores of bedrock; others are composed entirely of glacial debris.

Stratified Glacial Drift

Meltwater from the glacier carries off much of the finer debris and deposits it either on the beds of the overloaded streams (*glacio-fluvial deposits*) or in lakes (*glacio-lacustrine deposits*) and in seas. In places, the deposit may consist of *pitted outwash*, characterized by numerous undrained depressions ranging in size from mere dimples a few feet wide up to great holes a mile long and more than 100 feet deep. These holes are called *kettles*. They are formed by the melting of ice blocks that were detached from the main glacier as it wasted away, and were then partially or wholly buried in stream-deposited glacial debris.

The streams emerging from valley glaciers generally have somewhat steeper gradients than those from continental ice sheets. Nevertheless, they are so greatly overloaded that they immediately aggrade the valley floor to form a long narrow floodplain of coarse gravel and sand which grades downstream into finer sands and eventually into wide floodplains of silt. These aggraded valley bottoms, floored by glacio-fluvial debris, are called *valley trains*. When the glacier melts completely, the stream, being no longer overloaded, generally erodes the valley train into a series of terraces.

Stratified sand and gravel, containing here and there an irregular patch of till, forms long winding ridges in some areas from which continental glaciers have melted. Most of them are less than 100 feet high and a few hundred feet wide, but they may be several miles or even ten of miles long. They are called eskers, a term derived from the Gaelic meaning "ridge."

Eskers appear to be the deposits of aggrading streams that flowed in tunnels beneath the ice, or along the bottoms of crevasses, for some of them merge downstream into fans of outwash or into deltas. Most eskers must have formed when the ice sheet had wasted to a thin, almost stagnant mass, for, obviously, the strong thrust of an active glacier would close the tunnels and crevasses and scatter the esker deposit into ground moraine.

Where a glacier-fed stream enters a lake, a *delta* forms which may grow until it converts the lake to a swampy plain. The delta material is typically cross bedded and may range in grain size from coarse gravel to fine silt.

On the floors of glacial lakes, mud and silt accumulate as thinly stratified layers. Each layer is commonly only a fraction of an inch thick, giving the deposit a conspicuous paper like stratification.

Many glacial lakes are dammed on one side by the ice itself, and the water may extend into crevasses and irregular-shaped holes in the wasting glacier. Such lakes are generally temporary, and fluctuate in level as movement or melting of the ice diverts their outlets. Sediments accumulate in them. Because of fluctuations in lake level and movements of the glacier, these lake deposits are generally interlayered with stream deposits and with till. Upon melting of the glacier, patches of these sediments that filled former crevasses and holes in the ice are left in terraces and in isolated buttes called *kames* and *kame terraces*.

MODIFICATION OF TOPOGRAPHY BY GLACIERS

It is apparent that glaciers differ greatly from rivers in their mechanism of flow, and in their erosional and depositional features. The topography shaped by them contrasts sharply with the topography shaped by running water. Such differences are conspicuous in areas recently uncovered by glacial recession.

The polished and grooved bedrock floor is a feature not found in normal stream channels. Glaciated areas also contain swarms of lakes, ponds, and marshes that fill either shallow rock basins scooped out by the ice, or depressions dammed by moraines and outwash. The streams that connect these abundant lakes are totally ungraded. An immature stream pattern with numerous waterfalls and lakes is typical of recently glaciated areas.

The upstream sides of peaks, hills, or small rock knobs overriden by ice are commonly rounded, polished, and grooved; the downstream sides are irregularly jagged. It is inferred that the glacial rasp has sanded off the upstream side, but rock has been removed from the lee side, chiefly by quarrying of joint blocks caught by angular boulders dragged over them. Pressure relations may also be important. Pressure on the upstream face of the knob causes melting, and the water freezes in the low-pressure area, causing plucking on the lee side of the knob.

The usual mountain stream valley has a V-shaped cross profile but in valleys from which glaciers have melted the lower part of the profile is characteristically U-shaped, though remnants of a former V-shape may be preserved on surfaces higher than the top of the vanished glacier.

It is inferred that the spurs projecting between tributaries along the walls of a river canyon have been quarried away by the ice. A glacier cannot negotiate the sharp curves around spurs as easily as water; because of its bulk and brittleness it overrides the spurs and planes them into triangular facets forming a straight line of cliffs. At the same time, the floor of the valley is deepened by the glacial rasp. The widening, deepening, and straightening of former stream canyons by ice transforms the original V-profile to a U.

The long profile of a valley formerly occupied by a glacier is irregular and is commonly interrupted by abrupt steps above and below which the smooth U-shaped valley may continue with flat gradients. Several such "cyclopean steps" may appear in a single glaciated valley, alternating with polished rock floors and lake-filled basins.

Valley deepening by glaciers is not controlled by a smooth grade profile related to sea level as in streams, but by the thickness of the ice, the speed and duration of its flow, and the structure of the rock in its bed. In restricted parts of its course, it is not unusual for a valley glacier to deepen its trough far below sea level.

Unlike stream valleys, glaciated valleys commonly head in cirques. Abandoned cirques are conspicuous features in recently deglaciated mountains. These gigantic semicircular basins may be so closely spaced that the divides between them are reduced to knife-edged comblike ridges, or to triangular mountain "horns".

Stream valleys and glaciated valleys also differ in the relation of the tributaries to the main stream. The surface of a tributary glacier may accord with that of the main glacier—though, because of the rigidity of a thin ice mass, this is not always so—but the floor beneath the glacier and tributary is cut to depths depending on the volume and speed of the respective ice streams and the kind of rocks over which they move.

The thin tributary glacier, unless favored by less-resistant bedrock, cannot keep pace with the scouring and plucking by the larger glacier. As a result, when the ice retreats, the tributary valleys are common left "hanging" high above the floor of the main valley, and their streams descend into the main valley in high waterfalls or steep rapids.

Now that it is obvious glaciers leave clear marks of their former presence upon a landscape, let us see if there is evidence of widespread former glaciation in areas that today have a temperate or even a tropical climate.

Former Periods of Glaciation

The Iowa farmer, sweating under an August harvest sun, may doubt that "the present is the key to the past" if he is told that the soil he tills has come from deposits left by an ice sheet that once covered most of northern North America in the Pleistocene epoch.

Similarly, the dark-skinned native of the Talchirs in India, who pauses to rest from the steaming tropical heat on a polished and striated ledge of rock, will doubtless consider the idea fantastic, though alluring, if told that his perch is the floor of an ancient glacier that spread widely over India in Late Paleozoic time.

Conclusions such as these so tax the imagination that, in spite of the clearcut evidence, they were not accepted even by geologists until every possible alternative—from the Biblical flood to some mysterious kind of volcanic activity—had been tested and found inadequate to explain the facts recorded in the rocks and in the drainage patterns upon them. Yet, evidence of the presence of former ice sheets can be readily gathered by any interested observer.

THE PLEISTOCENE GLACIATIONS

In most of Scandinavia, southern Canada, Labrador, and parts of northern United States, the normal soil profile formed by weathering is either poorly developed or missing. Instead, rounded hilltops expose smoothly polished rock ledges like those beneath present glaciers. In places, the polish and smaller striations have been destroyed by weathering, but where protected by even a thin veneer of till, outwash, or peat, they may be as fresh and clear as beneath a modern ice sheet.

Boulders—many of huge size—are scattered erratically over the polished surface. Many boulders are of rocks entirely different from the local bedrock. In the fields of the Iowa farmer, boulders of gneiss and granite rest on flat-lying limestones and shales.

The headwaters of streams in the area expose no such bedrock, so these boulders could not have been brought in by floods. In parts of Iowa, chunks of copper are occasionally found like the copper ore mined from rock ledges oil the Keeweenaw peninsula in Michigan or Isle Royale in Lake Superior, but not found elsewhere.

Boulders of an unusual variety of granite called "rapikivi," known in outcrops only north of the Gulf of Finland, are scattered widely across Estonia and even far into Poland.

The basalt plateau of eastern Washington is strewn with huge granitic boulders of kinds found only many miles to the north. To reach their present position they must have been transported directly across the Columbia River canyon, which is here 1,500 to 2,000 feet deep.

A large nickel deposit in northern Finland was actually discovered by tracing its strewn ore fragments northward to their source near Petsamo.

Development of the Glacial Theory

The implications of such relations seem obvious to us now. Nevertheless, in the period from 1821-1835, when two European geologists, Venetz and Charpentier, showed that boulders of rocks characteristic of the central Alps were widely scattered across the broad Swiss plain to the north, and correctly inferred a former extension of the present Swiss glaciers, their ideas met little but skepticism.

In 1836, however, Charpentier induced one of these skeptics, a young Swiss geologist named J. L. R. Agassiz, to accompany him on a visit to the active glaciers and, lower in the valley, the huge abandoned moraines of the Rhone glacier.

Convinced on this expedition that the evidence for former more widespread glaciation was even stronger than Charpentier had claimed, Agassiz set to work on the problem.

He promptly saw and correctly interpreted the relation of the transported blocks to the polishing and grooving of the bedrock, and showed that these smooth striated forms could not possibly have been produced by water. Agassiz's work, too, met with general disbelief. As his biographer relates:

Men shut their eyes to the meaning of the unquestionable fact that ... the former track of the glaciers could be followed, mile after mile, by the rocks they had scored and the blocks they had dropped.

Agassiz, however, was not easily discouraged. In 1840, he visited much of Scotland, the north of England, and parts of Ireland. There, he demonstrated glacial phenomena identical with those on the Swiss plain, and he announced that not only had glaciers once existed in Britain, but that they formerly covered most of the country.

At first this conclusion raised a furore of objections, but it also set geologists to observing and more carefully analyzing the evidence. Faced with the facts so easily gathered over much of the British countryside, it was not long until Agassiz's views prevailed.

Agassiz emigrated to America and began the studies of glaciation in New England since followed so fruitfully by many investigators. Studies since Agassiz's time have supplied a wealth of information. Modern maps portray in detail the erosional forms and deposits of the former ice sheets. Many abandoned moraines have been traced in the field, lobe by lobe, and plotted on maps until their distribution in North America and Europe is well known. It is not difficult to map the moraines.

Although locally absent or poorly developed, most are conspicuous, easily identified, hummocky ridges of till that are readily traced for miles. Behind them, like signposts pointing the way, lie striated rock floors, locally strewn with till and foreign boulders. Here, too, are innumerable lakes and marshes where the vanished glacier gouged its floor or blocked older drainage by debris it left on melting. In front of the moraines are coalescing fans of outwash and, still farther beyond, sheets of loess, or terraces of silt.

All assist in locating the position of the edge of the vanished glacier, and all testify to its former existence. The landscapes of New York, New England, the Great Lakes states, and the Northwest are nearly everywhere stamped with these glacial imprints.

In mountainous areas, such as Yosemite National Park, the Cascade Mountains, or the Rockies, former glaciers are recorded by the numerous U-shaped valleys; the fresh moraines; the abundant cirques; the superb waterfalls pouring from "hanging valleys"; the clear mountain lakes nestled in rock-scoured basins; and the terraces of outwash gravel, sand, and silt extending downstream from the moraines.

Advance and Recession of Pleistocene Ice Sheets

Careful study of Pleistocene glacial deposits shows that they are not the result of a single advance and retreat of the ice. Instead, they record at least four stages of ice advance separated by long periods during which the glaciers withdrew to greater altitudes or latitudes than they occupy today, or even disappeared.

At many places in North America and Europe, roadcuts or natural exposures reveal two tills, one on top of the other. The upper layer of till may contain boulders of almost fresh granite and gneiss, some with polished facets and striae. In the lower till, however, beneath a layer of thoroughly weathered clayey soil, the outlines of similar boulders can also be made out, but the rocks have weathered to the consistency of cheese—the feldspars are thoroughly rotted to clay, and the ferromagnesian minerals have decomposed to limonite and other materials. Only chemically resistant rocks like quartzites are found intact in this lower till.

It is apparent that the older till had undergone long weathering and that typical A and B soil horizons had developed on it before the younger till covered it. The B-horizon of the weathered till is usually rich in clay that cements it into a tough sticky mass. Such tough clayey subsoils, whether of glacial origin or not, are popularly called *gumbo*. *Gumbotil* is the technical term used to designate deeply weathered clayey tills.

Moraines with mantles of gumbotil, unlike younger unweathered moraines, have generally been considerably eroded since they were formed.

Careful study of the degree of weathering of different glacial deposits and mapping of their mutual relations have permitted the discrimination of four main stages of ice advance. Between the ice advances the climate appears to have been mild and warm, by the fossils found in stream and marsh deposits between the tills.

DATING PLEISTOCENE DEPOSITS BY VARVED LAKE SEDIMENTS

Lake deposits are abundant in glaciated regions. Deltas grow rapidly into glacial lakes where torrents of turbid meltwater pour into them. The finest silt and mud settles slowly and is spread throughout the lake. Many glacial lakes are deep green, owing to the dispersion of light by the abundant particles of suspended clay.

In the winter, the lakes freeze over, and the small tributary streams may freeze solid. During this quiet period, the suspended clay particles beneath the ice settle slowly to the bottom. By spring, most glacial lakes have nearly cleansed themselves. In summer, however, the surface ice melts and the streams become gushing torrents of debris-charged water which sweep coarse as well as fine material out into deep water.

The coarse material settles promptly to the bottom, covering the fine winter deposit with a thin layer of coarse silt. Then, with the progress of the season, the silt is buried by successively finer deposits that gradually grade upward into the fine dark-colored mud of the next winter. In the summer, microscopic organisms flourish in the warm surface layers.

They extract calcareous matter from the water. The carbonaceous bodies of the organisms sink only slowly on death. Thus, the summer layer is richer in light-colored silt, the winter layer in black organic matter. Such cycles have been observed to form in many existing lakes.

PRE-PLEISTOCENE PERIODS OF GLACIATION

Many ancient sedimentary formations are composed of unsorted debris showing all the characteristics of till except that they are tightly cemented. They contain striated and faceted stones and are associated with varved shales and slates or with sandstones and conglomerates showing the typical features of outwash deposits. These features, and their local occurrence above polished and grooved rock floors, leave no doubt that they were formed during an ancient glacial epoch. Such cemented glacial tills have been called *tillites*.

Although small bodies of tillite have been found in rocks of many different ages, two periods in the earth's history before the Pleistocene are characterized by particularly widespread glacial deposits. During the Late Paleozoic, ice sheets spread widely over India, South Africa, Argentina and southern Brazil, southern Australia, the Boston Basin in the eastern United States, and many other localities.

In India, Australia, and South Africa, grooved rock floors covered by the hard tillite may be seen at hundreds of localities. Their preservation today is due to the fact that they were buried under a succession of younger sediments. Deep burial protected them from weathering and erosion, and the unconsolidated glacial deposits were slowly cemented into rock. Recent uplift and erosion of the overlying deposits has now exposed them to view.

In the Late pre-Cambrian also, glacial climates appear to have been widespread. Ancient tillites of this age have been found on every continent except South America.

CAUSES OF GLACIAL CLIMATES

Geologists and climatologists have tried for more than a century to explain the recurrence of glaciation on a continental scale. Theory after theory has been suggested, but all explain too little or too much. None can be considered satisfactory, at least in its present form; yet they have an interest that justifies at least brief mention.

Facts to Be Explained

- Continental glaciers (in Greenland and Antarctica) occupy about 10 per cent of the land surface today. They formerly covered about three times as much area during four different epochs of Pleistocene time.
- Climatic zones during the Pleistocene were about parallel to their present positions but were displaced to the south during the glacial maxima and to the north during interglacial epochs. In low latitudes high rainfall was contemporaneous with glacial climates at higher latitudes.
- Estimates of the duration of the several glacial and interglacial stages of Pleistocene time do not suggest periodic recurrence; the climatic fluctuations seem to have been irregular.
- Evidence of continental glaciation on scales approaching that of the Pleistocene is also found in Late Paleozoic rocks (about 200 million years old) and in pre-Cambrian rocks (at least 500 million years old), but not in strata of intermediate age.

Suggested Explanations

The many theories that have been offered may be grouped under the following general categories: geographic, atmospheric, and oceanic; geophysical; astronomical; and cosmic.

The geographic theories attempt to account for glaciation by climatic changes that would be expected from changes in the areas of the continents with respect to the oceans, changes in oceanic and atmospheric circulation due to changes in the shape of ocean basins, mountain uplifts, and so forth. These causes are inadequate because there is no evidence whatever of any notable shift in the distribution of landmasses or of violent fluctuations in their altitude since Pliocene time, yet we have conclusive evidence of four great glacial advances and of milder-than-present climates in some, at least, of the interglacial times.

It has been suggested that changes in the amount of carbon dioxide and of volcanic dust in the atmosphere would bring about climatic variation. Carbon dioxide absorbs some of the heat radiated to space from the earth's surface. If its quantity were greater, more heat might be "blanketed in" and the temperature would rise, whereas, if it were less, the climate would become colder. However, this mechanism is quantitatively inadequate, especially because the effect of simultaneous change in amount of water vapour would practically compensate for the variations in carbon dioxide.

Volcanic dust undoubtedly screens out some of the sun's radiation, but there is no sign that volcanoes were more active during the stages of ice advance than during interglacial times. Changes in salinity of sea water (and consequently modifications of the oceanic currents and their climatic influences) are likely results from glaciation but they can hardly have brought it about.

Geophysical theories that attribute glaciation to shifts in the positions of the continents are obviously not applicable to the Pleistocene climatic fluctuations, nor is there any strong evidence that these could ever have sufficed for earlier glaciations. Although it is conceivable that the amount of heat given off from the interior of the earth may have fluctuated from time to time, measured fluctuations are so small compared to the heat received from the sun (less than 1/10,000) that they seem negligible as a climatic factor.

The astronomical theories take account of the periodic variations of the earth's motion with respect to the sun. There are several such variations and there can be no doubt that they would have some climatic effect. Each affects the distance of a particular point on the earth's surface from the sun and hence the amount of solar radiation it receives. They could not have alone produced the glacial periods, however, because their effects on northern and southern hemispheres, though not always exactly opposed, would not be identical, either. Yet, we now have recession of glaciers taking place simultaneously in both hemispheres and it is reasonable to think that earlier fluctuations were also simultaneous.

Furthermore, since all these astronomic motions have been going on through geologic time, we should have had continental glaciation every few hundred thousand years. Instead, we must go back to the Permian to find anything comparable to the Pleistocene glaciation in the geologic record. These geologic arguments seem conclusive, but several meteorologists contend, in addition, that the temperature changes produced by this mechanism would be entirely too small to produce glaciation even in one hemisphere.

The cosmic theories are of two kinds:

- That the solar system from time to time encounters cosmic dust clouds.

- That the sun varies in the amount of heat it radiates. The cosmic dust clouds may either screen out the sun's radiation (if the component particles are fine enough) or blanket the earth's radiation to outer space (if the particles are somewhat larger). This is, of course, a very flexible hypothesis, as there is no way of testing it; with one assumption we can get a colder climate, with another a warmer.

Accordingly, this idea is not seriously considered by most geologists, even though it may be the true explanation, for the scientific approach is to consider seriously only those hypotheses that are susceptible of testing in some fashion, however indirectly. The second suggestion, that the sun's radiation fluctuates somewhat over long periods of time, is on a better footing.

The sun's radiation does, in fact, fluctuate, and with it the amount of heat received by the earth. Variable cloudiness and other atmospheric factors make accurate measurements difficult, but short-term variations of as much as 3 per cent in the heat output seem well established.

The American astronomer, Abbot, a leader in this research, has declared that larger solar changes may well have taken place in the past, even though it is not yet possible to prove it. Several climatologists have tried to deduce the consequences of such moderately large changes, if they do, in fact, occur. We have not space to review all the hypotheses that have been suggested; perhaps the one that has won most support from meteorologists is due to the British meteorologist, Sir George Simpson.

Simpson (1934) reasoned that if solar radiation increased, the air temperature would rise all over the earth, winds would increase and so would cloudiness and precipitation. The increased precipitation would lead to greater accumulation of snow and the increased cloudiness would lessen summer melting.

The ice caps in both hemispheres would expand. However, the rising temperature would eventually prevent further advance of ice and ultimately cause melting.

At the high point on the radiation curve, the world climate would be milder and wetter (more "oceanic") than it is today. When radiation again decreased, the reverse sequence would be followed: appearance and advance of the ice; then retreat because of lower precipitation; and, finally, when radiation fell to its present value we would be back to present conditions, with polar accumulations of relatively small size. Thus, the somewhat paradoxical condition would arise, that a rise in the general temperature of the air would, by increasing precipitation and cloudiness, lead to continental glaciation of much of the present temperate zone.

On this theory, one rise and fall of radiation would produce two glaciations in high latitudes, separated by a warm and wet interglacial time. The climate following the second glacial would be cold and dry like the present. In low latitudes there would be a single epoch of increased rains that would endure through both high-latitude glacials and the intervening warm, wet interglacial.

The four Pleistocene glaciations recognized in Europe and North America would, on this hypothesis, require two cycles of increased solar radiation, separated by a time during which the climate may have been much as it is today. Simpson computed that, during the maximum of solar radiation, the mean temperature was 5° to 10° C. warmer than at present and that the average cloudiness was increased from the present 54 per cent to between 70 and 80 per cent. Simpson computed that these changes are reasonably to be expected if the sun's radiation fluctuated about 20 per cent above its present value.

This hypothesis is attractive because it would account for the increased precipitation that would be necessary to feed ice sheets. The present line of mean annual temperature at the freezing point lies far closer to the equator than the limit of present ice sheets, and this clearly indicates that low temperature alone is not enough to produce glaciation but that increased precipitation is needed.

With the present configuration of land and sea, which is not greatly different from that during the Pleistocene, increased precipitation can only be brought about by increased evaporation from the oceans. The theory also accounts for the warm, moist climate suggested by plant fossils from some interglacial deposits. On the other hand, there is no general evidence that the last interglacial period was generally warmer and moister than the second, as it should have been on this hypothesis. Nor is there general agreement that the increased rainfall in lower latitudes lasted throughout both the last two glacial epochs and the intervening interglacial epoch.

Finally, inasmuch as most astronomical processes are cyclic, if the variation of the sun's radiation is also cyclic, we should have had numerous other glacial episodes throughout geologic history. We must conclude that, even if this theory is satisfactory in some respects, an auxiliary hypothesis is needed to account for the long nonglacial periods of the geologic record.

We thus see that no satisfactory hypothesis has yet been advanced to explain the widespread glaciation of the continents during the Pleistocene. It is impossible to say whether or not the present is simply another interglacial epoch. We cannot tell whether glaciers will again encroach on the sites of Chicago, Copenhagen, and Warsaw within a few thousand years, or whether the climate will remain about the same as it is today. Much more work will have to be done to win an answer to this fascinating problem.

THE EFFECT OF GLACIAL LOADS ON THE EARTH'S CRUST

There is strong evidence from deflection of the plumb line and from gravity measurements that large segments of the earth's crust are essentially in isostatic balance. Glaciers of continental dimensions also afford clear geologic evidence in support of this conclusion.

The fact that former glaciers flowed over and eroded the ground on which they rested is proof that their surfaces sloped down from their feeding grounds to their termini. Wherever the glaciers spilled around mountainous topography, it is possible to get some measure of their thickness. In New England, for example, the highest peaks were unglaciated, and a measure of more than 4,000 feet has been made for the Pleistocene ice in that region. In southern Canada the ice would certainly have been considerably thicker.

Although the density of ice is only a little more than one-third that of ordinary rock, a load of 3,000 feet of ice over an area as large as the Great Lakes and southern Canada would be the equivalent of a load of more than 1,000 feet of rock placed on the area, and should bend the crust down.

Though the isostatic response by subcrustal flow might take considerable time, owing to the high viscosity of the subcrustal material, the lowering should be considerable and measurable, if isostasy really is general. Fortunately, the shorelines of ancient glacial lakes give us a means of testing the idea.

They were level surfaces when formed. After the ice load melted away the rise of the crust in response to unloading should tilt the shorelines, causing them to rise to the north. This is exactly what is found, not only in the Great Lakes region, but in the Baltic, in Labrador, and even in the basin of Lake Bonneville, where the load was only 1,000 feet of water covering a much smaller area than that of the great continental glaciers.

Suffice it to say here that the testing of the isostatic principle by glaciers is in full agreement with the geodetic evidence from which the theory was first announced.

Chapter 10

Mountain Ranges and Ocean Basins

One of the unexpected discoveries in earth science in the previous century was that of a fundamental difference between continents and ocean basins. Ocean basins are not merely the low lying parts of the Earth's surface flooded by salt water but are great, relatively steep-sided, structural depressions. In fact, there is too much water for the size of the ocean basins, and parts of the continents are now flooded and probably have been flooded through a great deal of the geologic past.

Precise measurements of gravitation attraction in major mountain ranges, continental areas, and over the ocean basins showed an even more unexpected feature. The continents and mountain ranges do not represent extra loads of rock superimposed upon the Earth's crust, but are masses of lighter rock floating in a denser substrate. An iceberg floats above the water much in the same fashion, buoyed up by deep submerged roots.

The great mountain ranges of the world and the continental masses similarly have deep roots of light rock penetrating down into the denser crust, and thus the mountain ranges and continents float at elevations appropriate to the depth and size of these submerged light roots.

Thus, all the major features of relief of the surface of the Earth show mirror image features within the crust. The major mountain ranges float at high elevations because they are buoyed up by light rocks. The continents float at intermediate elevations with roots of intermediate depth, and the deep oceans are underlain by thin layers of light rock.

Seismologists, from the study of earthquake waves, have shown that the Earth's mantle is solid to the depth of the outer core, some 2900 kilometers. The observation that large mountain ranges and continental masses float on the crust of the Earth at elevations appropriate to the size and density of their roots implies that rocks at shallow depths in the Earth's mantle, although solid, have little strength and can flow in response to small stresses given sufficient time.

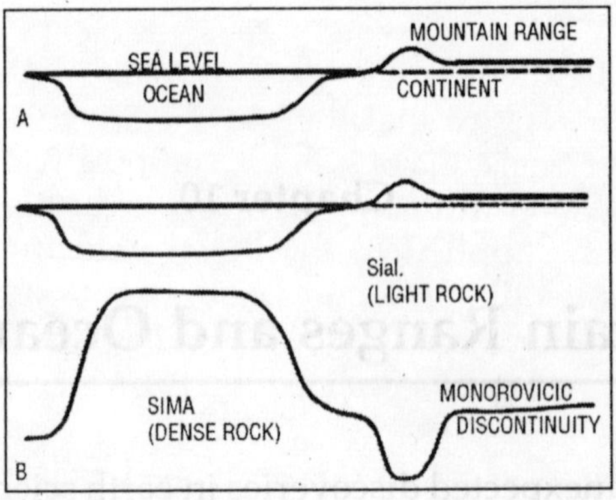

Fig. Profiles Through the Earth's Crust

This deduction is strengthened by the observation that rocks in deep, eroded, old mountain chains are intensely contorted and folded, plain evidence that, at high pressures, solid rocks can flow readily and do not have great strength.

Recognition that continental rocks are lighter and more buoyant than oceanic rocks gave rise to the concept that the crust of the Earth is made of two contrasting materials: sialic material, rich in silicon, the alkalies and aluminum, making up the continents; and simatic material, richer in iron and magnesium, making up the denser rocks below the floor of the ocean and lying under the sial of the continents. The sial is assumed to be granite or granodiorite in composition, and the sima is assumed to be basaltic in composition.

Early in this century, the Yugoslav seismologist, Mohorovicic, obtained evidence from seismograms that earthquake waves, traveling a few tens of kilometers below the surface of the Earth, gave records showing sharply higher speeds for both shear and compressional waves than earthquake waves traveling near the surface.

This indicated an abrupt change in rock types at a few tens of kilometers under the continents and at a few kilometers under the oceans. In recent years, extensive studies have produced a fairly clear general picture of the nature and depth of this level of change, or discontinuity, under the continents, mountain ranges, and ocean basins.

This discontinuity, called the Mohorovicic or M discontinuity, is at a general depth of 30 to 40 kilometers under the continents, but may be as deep as 60 kilometers under the roots of major mountain chains. It is as shallow as 4 to 5 kilometers below the floor of the deeper parts of the ocean. The discovery of the M discontinuity seemed to confirm the notion that the crust is fundamentally made up of two different kinds of rock material. The discontinuity itself appears to be the boundary between these, the sialic rocks above and the simatic rocks below.

The rocks below the M discontinuity have seismic velocities and densities which suggest that they may be even richer in magnesium and iron than normal sima of basaltic composition. Consequently, they are, by some, called ultrasima. However, throughout this paper the word sial will be applied to the lower velocity rocks above the M discontinuity, including the range of basalts to granites, and the word sima will be applied to the denser rocks below the M discontinuity. Prior to and along this general picture, the concept of isostasy developed.

This is the notion, previously discussed, that the lighter continental rocks float at an appropriate depth, depending on their mass and mean density, in a denser substratum. As rock is eroded from the tops of continents and mountain ranges they tend to float up higher and higher, renewing their relief, permitting erosion to continue. Four facts, however, sharply contradict this picture of a sialic crust of varying thickness floating on a simatic substratum of different chemical composition and different density.

1. Large areas of continents, long near sea level, have been uplifted many thousands of feet in the air. Further, this uplift seems to have taken place rather rapidly in terms of geologic time.
2. Sediments of low density, filling troughs along the margins of continents, apparently are able to subside into this higher density substratum.
3. Inasmuch as radioactive, heat-producing elements are associated with sialic rocks, one might expect heat flow through the thicker parts of the Earth's crust to be much greater than through the thinner parts of the Earth's crust. However, as a first approximation, heat flow through the crust of the Earth is approximately the same through continents, mountain ranges, and ocean basins.
4. The lifetime of continents and mountain ranges is vastly greater than rates of erosion would suggest.

Let us examine each of these apparent facts and their consequences on the hypothesis of sialic continents floating on a simatic basin. The problem of the uplift of large plateau areas is one which has puzzled students of the Earth's crust for a very long time. Regions which are at sea level, or near sea level, may, over a relatively short geologic time span, such as a few million years, be uplifted several thousands of feet.

The Colorado plateau and adjacent highlands is an example. Here, in an area of approximately 250,000 square miles that apparently stood at sea level for several hundreds of millions of years was uplifted approximately one mile vertically some 40,000,000 years ago in early mid-Tertiary time and is still a high plateau. The Grand Canyon of the Colorado has been carved through this great uplifted plateau.

Given an Earth with sialic continents floating in a denser simatic substratum, what mechanism would cause a large volume of low standing continents to rise rapidly a mile in the air? Furthermore, evidence from gravity surveys suggest that the rocks underlying the Colorado plateau are in isostatic balance, that is, this large area is floating at its correct elevation in view of its mass and density. Recent seismic evidence confirms this, in that the depth to the M discontinuity under the Colorado plateau is approximately 10 kilometers greater than over most of continental North America.

Thus, appropriate roots of light rock extend into the dense substratum to account for the higher elevation of the Colorado plateau. We have then a double-ended mystery, for the Colorado plateau seems to have grown downward at the same time that its emerged part rose upward. This is just as startling as it would be to see a floating cork suddenly rise and float a half inch higher in a pan of water.

To date, the only hypothesis to explain the upward motion of large regions like the Colorado plateau is that of convection currents. Slowly moving convection currents in the solid rock, some 40 to 50 kilometers below the surface of the Earth, are presumed to have swept a great volume of light rock from some unidentified place and to have deposited it underneath the Colorado plateau.

A total volume of approximately 2,500,000 cubic miles of sialic rock is necessary to account for the uplift of the Colorado plateau. While it is not hard to visualize rocks as having no great strength at the high pressures and temperatures existing at depths of 40 to 50 kilometers, it is quite another matter to visualize currents in solid rock of sufficient magnitude to bring in and deposit this quantity of light material in a relatively uniform layer underneath the entire Colorado plateau region.

The Tibetan plateaus present a similar problem, but on a vastly larger scale. There, an area of 750,000 square miles has been uplifted from approximately sea level to a mean elevation of roughly three miles, and the Himalayan mountain chain bordering this region has floated upward some five miles, and rather late in geologic time, probably within the last 20,000,000 years.

The quantity of light rock which would need to be swept underneath these plateaus by convection currents to produce the effects noted would be an order of magnitude greater than that needed to uplift the Colorado plateau, that is approximately 25,000,000 cubic miles.

Even more troublesome than the method of transporting all this light rock at shallow depths below the surface of the Earth is the problem of its source. The region from which the light rock was moved should have experienced spectacular subsidence, but no giant neighboring depressions are known. A lesser but large problem is how such enormous quantities of light rock can be dispersed so uniformly over so large an area.

This evidence of uplift and down sinking of various crustal blocks, with the blocks always remaining in approximate isostatic balance, does not seem to harmonize with the view of a floating sialic continent on a denser substratum where one might expect to find little variation in elevation with time.

The second problem, that of the subsidence of troughs, is of equal difficulty. The rivers of the world carry enormous quantities of sediments seaward. Most of this sedimentary burden is deposited within a few tens or hundreds of kilometers of the shore line and little is transported to the deep ocean basins. Thus, elongate prisms of sediments are built up parallel to the shores of certain regions where great quantities of sediments are transported to the sea.

The crust, in response to this added load of sediments, begins to buckle downward. Troughs filled with sediments appear, paralleling the coast line. The chicken and the egg argument enters here, for it is not entirely clear whether deposition of sediments generates the troughs or whether the troughs are formed first and are later filled with sediments.

However this may be, one such trough now in the making is along the coast of the Gulf of Mexico on both sides of the mouth of the Mississippi River. Surprisingly enough, this trough deepens at about the rate new sediments are added to it. Thus, the sediments are always deposited in relatively shallow water.

Fundamental laws of physics are violated and on a large scale if this down warping is produced directly by continued loading of sediments. These deep troughs filled with sediments may contain 50,000 to 100,000 feet of sediments and may be 1000 or more miles long and 100 miles in width. The mean density of the sediments, even compacted under a load of 10,000 feet of other sediments, is approximately 2.4 to 2.5.

The rocks displaced in the down warping trough are known to be denser, with a mean density of 2.8 to 2.9. By what mechanism do light sediments displace denser, crystalline rock? These troughs of sediments, like the plateaus considered earlier, always appear to be in isostatic balance. If the conventional is to be sustained, dense rock must automatically be removed from below the bottoms of these sedimentary troughs at approximately the same rate that they receive sediments from the rivers which feed them so that the troughs balance and float with their upper layers of sediments under a few tens or hundreds of feet of water most of the time.

The problem of the mechanics of the formation of deep troughs of low density sediments is heightened when their full history is considered. Many are known in the geologic record. In most, sediments accumulate for perhaps a hundred million years and reach a total thickness of as much as 100,000 feet. These thick, highly elongate lenses of sediments may then be slowly folded and uplifted to form mountain ranges which may initially stand as much as 20,000 feet high.

Surprisingly, the geologic record shows that a large fraction of the mountain ranges of the world have been formed from rocks of these thick, geosynclinal troughs. Extensive volcanic activity may accompany and continue beyond the time of the formation of the mountain ranges.

The mystery, then, of the down sinking of the sedimentary troughs, in which low density sediments apparently displace higher density rocks, is heightened when we note that these narrow elongate zones in the Earth's crust, down warped the most, with the greatest accumulation of rock debris, shed by the higher portions of the continents, become in turn the mountain ranges and the highest portions of the continents.

The third of the major problems connected with the postulated sialic continental area and simatic oceanic region is that pointed out by recent measurements of flow of heat through the crust of the Earth.

A considerable number of recent measurements have been made of temperature gradients and rock conductivities within the outer part of the Earth's crust. Careful temperature profiles have been made within many of the accessible deep mines and in numerous wells and tunnels. From these data, a fairly reliable picture has developed of heat flow within the Earth's outer crust, although measurements are not nearly so detailed or as numerous as is to be desired.

The rate of escape of heat through most continental areas appears to be approximately 1.2 microcalories per centimeter per second. It has been known for many years that most of the heat escaping from the Earth is radiogenic heat, generated in the Earth by decay of radioactive isotopes of uranium, thorium, and potassium. Little or none of the heat escaping from the Earth is primary heat, inherited from an initially hot Earth.

In fact, there is no compelling evidence that the Earth was molten in its youth or even formed from hot material. We know that the rocks near surface today appear to be in fairly reasonable thermal balance. The rate of heat escaping from them to the surface of the Earth is very close to the rate at which heat is generated in them by radioactive decay of certain elements.

Over the last twenty years, extensive data have been accumulated concerning the distribution of the radioactive elements. Uranium, thorium, and potassium are 10 to 100-fold as abundant in the light silica-rich rocks as they are in denser simatic material, rich in magnesium and iron, and low in silica. Consequently, we might expect that radiogenic heat in the thick sialic continents should be vastly greater than the heat generated in the presumably radioactive-poor simatic material underlying the ocean floors.

Further, we would expect heat flow to be greatest in the thickest parts of the continents, that is, in mountainous regions buoyed up by thick roots of sial rich in radioactive elements. A number of studies of heat flow through the continents have been made over the last two decades. These studies have been made by examining the distribution of temperatures and rock conductivities down deep wells and along tunnels.

Surprisingly, these studies show almost no correlation between mean elevation of land mass and heat flow through the Earth's crust. This was most unexpected because all the broader regions of higher elevation are presumably underlain by thick zones of light rock which, from all determinations, should be richer in radioactive elements.

Nonetheless, it was confidently expected that heat flow through the floor of the ocean would be a fraction of that observed in the continental land masses. The first measurements of heat flow through the floors of the ocean were reported in 1952 by Sir Edward Bullard. These determinations of heat flow were ingeniously made by inserting probes containing thermisters into the muds on the floors of the oceans.

Startlingly, the heat flow determined by these measurements through the floor of the ocean was almost identical with that measured in continental and mountainous areas. Later results by Revelle and Maxwell, although indicating wide ranges of heat flow from place to place in the oceans, have only affirmed the earlier observation that heat flow through the ocean floor is essentially the same as that on the continents.

There seem to be only two possible explanations for this most unexpected discovery: either the concentration of radioactive elements in the rocks below the floor of the ocean is the same as that in rocks which make up the continents or else heat is transferred by some special mechanism from deeper down in the Earth to near-surface sites underneath the oceans. If the concentration of radioactive materials in the few tens of miles below the floor of the ocean is the same as in a few tens of miles below the continents, then our previous view that the floors of the oceans are underlined by radioactive-poor sima and the continents were underlain by radioactive sial certainly cannot be right.

The alternative explanation, equally difficult, is that high temperature rocks from deeper down in the Earth are convectively carried up to near-surface environments below the oceans. Thus, heat escape through the floor of radioactive-poor oceans is fortuitously approximately the same as heat escape through the radioactive-rich continents.

The fourth problem, that of the long lifetime of continents and mountain ranges, is perhaps the most difficult of all. The rivers of the world strip tremendous quantities of rock debris off the continents each year and deposit it in the oceans. The Mississippi, for example, contains about onehalf weight per cent of solids as it flows into the Gulf of Mexico.

Each year, it brings to the Gulf of Mexico approximately 750 million tons of dissolved and solid material. The great rivers are steadily wearing down their basins. Calculations show that the Missouri River lowers its drainage basin about one foot in each eight thousand years, and that the rate of erosion for the entire United States approximates one foot in 10,000 years. At this rate, all the land masses of the world wouldbe eroded to sea level in something of the order of 10-25 million years.

This is particularly surprising in view of the fossil record. Land animals and plants have been known on the surface of the Earth for well over 300 million years, and the sedimentary record indicates high land masses extending back at least two billion years. Much geological evidence indicates that the ancient continents were in approximately the same place as the present continents and that continents have existed more or less as they are today and for a period of at least two billion years.

How do we reconcile an erosional lifetime for continents of something like 25 million years with a known lifetime of something of the order of two billion? Why has not all the continental sial been uniformly distributed through the ocean basins?

The mountain ranges bordering the continents and interior to the continents present an even more difficult problem. The rates of erosion along the slopes of steep mountains are many times those of lower lying continental land masses.

The lifetime of mountains, therefore, must be far less than the 25 million years estimated for continents. In contrast to this reasoning, however, is the geologic record which strongly suggests that the Appalachian Mountain Range has existed more or less where it is today and, as far as we know, with reasonably similar relief for the last 200 million years, shedding sediments both to interior valleys and coastward.

Thus, we see orders of magnitude discrepancy between lifetimes of mountain ranges and continents, estimated on the basis of known rates of erosion, and the lifetimes of the mountains and continents as indicated by the geologic record. Even though we assume that mountain ranges and continents are somewhat analogous to icebergs that float up as their exposed portions are melted away, the presumed depth of roots of the mountain ranges and thicknesses of the light continental rocks permit extension of the estimated lifetime of continents by no more than tenfold that based on present erosion rates and mean elevations.

Thus again, the notion that the rocks which make up the continents are grossly different in composition from those underlying the ocean basin does not seem to hold up, for we would expect that the rain waters washing over the continents would have long ago dispersed the continental rocks into the oceans.

These four major observations then — persistence of continents and mountain ranges in spite of high erosion rates, the relatively uniform values for heat flow in continents and ocean basins, subsidence of marginal troughs in response to loading by low density sediments, and uplift of plateaus once worn to sea level — suggest the inadequacy of the traditional view that continents represent masses of low density silica and aluminarich rock floating in the denser media of sima, iron, and magnesium-rich rock.

Recent theoretical studies by Gordon J. F. MacDonald and experimental work by Robertson, Birch, and MacDonald and by the writer, as well as interpretation by J. F. Lovering, suggest a different structure of continents, a structure which simultaneously explains most of the observed phenomena associated with continents, mountain ranges, and ocean basins and accounts for the four major stumbling blocks in existing theory. This new model of the Earth's crust stems from theoretical considerations largely confirmed by recent experimental work in the field of high pressures.

Very many crystalline solids undergo polymorphic inversions to denser phases when subjected to high pressures. The behaviour of matter at high pressures has been extensively investigated by P. W.

Bridgman who has demonstrated literally hundreds of polymorphic inversions among common substances in the pressure range 0-100,000 atmospheres.

Graphite and diamond form, for example, a familiar polymorphic pair. At sufficiently high pressures and temperatures graphite may be converted to diamond. A temperature of 1500 K and a pressure of 100,000 atmospheres is sufficient for the conversion, and, indeed, many thousands of carats of diamonds are now being made annually by General Electric Company by subjecting carbonaceous material to high temperatures and pressures.

It has long been noted that basalts and eclogites, rocks with sharply contrasting mineralogy, have essentially identical chemical composition.

	Eclogite	*Plateau Basalt*
SiO_2	48.12	48.80
TiO_2	.85	2.19
Al_2O_3	10.42	13.98
CaO	9.99	9.38
MgO	14.22	6.70
FeO	13.92	13.60
Na_2O	1.45	2.59
K_2O	.58	.69

Eclogite, however, contains no feldspar; instead, it is made up of jadeitic pyroxene and garnet. The mean density of eclogite is 3.3 gm/cc, that of gabbro is 2.95 gm/cc. As eclogite is the denser of the two phase assemblages, it is the rock which must exist at the higher pressures.

The density contrast, about 10%, between gabbro and eclogite is almost the same density contrast believed from seismic evidence to exist at the M discontinuity, although the contrast at the discontinuity has usually been assumed to be a chemical contrast rather than a phase contrast. Indeed, Fermor, Holmes, and Goldschmidt suggested that M discontinuity might be a phase contrast and that the rocks below it are eclogite.

Their suggestion received little discussion or acceptance but has been recently revised by G. J. F. MacDonald on the basis of calculations of the pressure-temperature conditions controlling the phase change of nepheline plus albite to jade and of albite to jade plus quartz.

These thermochemical calculations have been confirmed by experimental work of Robertson, Birch, and MacDonald and by the writer. These two experimental studies, though in disagreement in detail, confirm the calculations based on thermochemical data that, at pressures of 15,000 to 25,000 atmospheres, depending on temperature, the nepheline plus albite undergoes a polymorphic change to jade, and albite undergoes an inversion to jade plus quartz at slightly higher pressures.

Further, an experiment made by me on basalt glass showed that, at 500° and pressures below 10,000 bars, basalt glass crystallized as gabbro. The major mineral component is feldspar. At pressures above 10 kilobars and at a temperature of 500°, the amount of feldspar decreases and, finally, basalt glass crystallizes directly to a rock made up dominantly of jadeitic pyroxene. Identification of phases were by X-ray. Significantly, 500° and 10 kilobars are approximately the temperatures and pressures estimated at the M discontinuity underneath the continents.

It thus appears that the M discontinuity may reflect a phase change from gabbro to eclogite rather than a change in chemical composition. This phase change will account for the observed change in seismic velocity from approximately 6.5 kilometers per second to 8.1 kilometers per second and a change in density from 2.9 to approximately 3.23. Thus, the older suggestions of Fermor, Holmes, and Goldschmidt are supported by field measurements, theoretical calculations, and recent experimental work.

If the discontinuity caused by a phase change takes place at a depth of 30 kilometers, a depth equivalent to a pressure of approximately 10 kilobars under the continent, how do we account for the much greater depth to the discontinuity under mountain ranges and the much shallower depth to the discontinuity under the oceans?

The answer lies in the fact that the change takes place at a different pressure for a different temperature. As near as can be told from the computations and from the experimental data, the slope of this phase change is approximately the same as the Earth's pressure-temperature gradient.

Consequently, if it is assumed that the Earth's temperature increases a little more rapidly per foot of depth under mountain ranges than under continents generally, the transition will take place at a vastly greater depth. If it is assumed that the Earth's temperature increases with depth a little more slowly under the oceans than under the mountains and continents, the transition is at shallow depths, the single transition explains the varying depths to the M discontinuity under the oceans, mountain ranges, and continents.

We assume that there are variations in temperature from

continents to ocean basins to mountain ranges, and, consequently, we would expect variations in heat flow. However, the necessary variations in heat flow to account for these different depths of intersection are exceedingly small, well within the range of observations and are certainly not the threefold variations in heat flow that we would expect if the continents and mountain ranges were thick zones of radioactive-rich sial and the ocean was underlain by radioactive-poor sima.

The Earth's pressure-temperature gradient is almost the same under the oceans as is the slope of the phase change. The intersection here is assumed to be at low pressures and temperatures. Because the temperature is very low, reaction rate of the phase change might be expected to be very slow and the response of the discontinuity position under the oceans might be extremely sluggish to small changes in temperature and pressure. Thus, we may not always have thermodynamic equilibrium under the oceans.

Early in this discussion it was noted that the relief of the Earth's crust is a direct function of the thickness of the zone of light rock. If the thickness of the zone of light rock reflects the depth to the M discontinuity, which it almost certainly does, the relief of the Earth's crust can be interpreted as mirroring the various temperature gradients in the upper part of the mantle.

The four major problems of the surface of the Earth, discussed earlier, seem satisfactorily explained by phase transition. A chemical contrast at the discontinuity is unnecessary. The rocks on both sides of the M discontinuity may thus be of the same composition and the depth to the discontinuity may be a function of very slight temperature variations from place to place in the Earth's crust.

The uplift of continents, once at sea level, to high plateaus would be a consequence of warming the rocks near the M discontinuity a few tens of degrees. When this happened, the phase change would migrate downward to much greater depths. The dense rock below the discontinuity would become light rock and the volume increment would float the continents to higher levels. Thus, convection currents are no longer needed to transport millions of cubic miles of light material underneath the continents in order to float them higher into the air.

Similarly, the long lives of mountain ranges are explained. As the tops of mountains are eroded away, pressure at the discontinuity deep below the mountains decreases. Dense rock at the discontinuity would be converted to light rock, so light roots underneath the mountains would be recreated to keep them floated to high elevations.

The down sinking of sediments in troughs is also explained by the phase transition. If sediments from a mountain range were rapidly removed and deposited in troughs, the first effect of loading would be to increase the pressure at the base of the trough with very little change in the temperature. Consequently, the discontinuity would migrate toward the surface.

The trough would sink, not only because of the added load of rock at the surface, but because light rock would be converted into dense rock at the discontinuity below the trough with a consequent decrement in volume of material below. Thus, the short-time effect of rapid sedimentation is one of sinking. A most interesting long-time effect appears. The added new sediments filling the trough are of low thermoconductivity and possibly richer in radioactive material than the surrounding rock.

Consequently, given sufficient time, the temperature would slowly rise at the bottom of the trough, and, although the discontinuity would first migrate surface ward under response to loading, it would ultimately migrate downward under response to the rise in temperature owing to the blanket of poorly conducting sediments rich in radioactive elements deposited in the trough. Thus, troughs might sink for considerable time and then be uplifted to form mountain ranges as the roots of the trough deepen with warming of the base.

This implies that mountains are generated largely because of vertical motion and not lateral thrust. A good deal of the faulting and folding of rocks in mountain ranges is assumed to be the result of load. By this thesis, the major folds and faults associated with mountain chains are gravitational in origin, though concomitant lateral thrust of other origin is not excluded.

The problem of the relatively uniform heat flow to the surface of the Earth is readily explained by the phase transition concept. The earlier crustal models assumed continents were made up of silica-rich and radioactive element-rich rocks. Thus, continents should, but do not, show heat flows several times that of oceanic areas. If the bulk composition of continental rocks were not vastly different from the bulk composition of oceanic rocks, we would expect relatively uniform heat flew from place to place in the Earth's crust.

This is indeed what we do find. The precision, however, of measuring heat flow is not sufficiently great to exclude the possibility that minor variations in temperature do exist from place to place in the Earth's crust. In fact, it is necessary to appeal to these minor variations to account for the existence of ocean basins, mountain ranges and continents on the basis of a phase change as discussed here.

If we assume the M discontinuity to be a phase change, many questions are answered, but other questions are also raised. The phase change cannot be a simple solid-solid phase change inasmuch as the major minerals involved are of variable composition. Consequently, the change must take place over a considerable depth interval and should not be a sharp change taking place at a fixed depth. The data of seismology bear on this problem.

They permit the interpretation that the discontinuity may take place, instead of at a given depth, over an interval of as much as 10 kilometers under the continents. This is within the requirements of the change. However, more difficult problems emerge when oceanic areas are considered. The discontinuity under the oceans is very shallow and apparently takes place over a very narrow depth interval. In fact, the pressure interval seems much too narrow for it to represent the gabbro-eclogite change.

However, further experimental work needs to be done to measure precisely the required pressure interval and more refined seismic work will be necessary before we know exactly the distribution of seismic velocities below both the oceans and the continents.

DEVELOPMENT OF THE HYDROSPHERE AND ATMOSPHERE

Abstract: A satisfactory hypothesis of the development of the hydrosphere and atmosphere depends upon evidence from many sciences and the solution of many other fundamental problems of earth history.

But because it is so closely related to many other problems, any progress toward' unraveling the history of the hydrosphere and atmosphere limits the range of permissible speculation about such distantly related questions as the origin of the solar system, continents, mountains, and living organisms.

Several hypotheses of the source of the earth's air and waters are examined for their consistency with established principles and observed geologic evidence, and special attention is given to the probable composition of the early atmosphere.

Hypotheses of the origin of the atmosphere and hydrosphere fall into two chief categories:

- That all air and water of the earth are residual from a dense primitive atmosphere that once enveloped a molten globe.
- That they have accumulated at the earth's surface by leakage from the interior.

The quantities of water, carbon dioxide, organic carbon, nitrogen, sulfur, etc., that have been or are now part of the earth's atmosphere and hydrosphere may be estimated within reasonable limits of uncertainty and these "excess" volatiles afford a basis for testing chemical consequences of the alternative hypotheses. Several writers have suggested that the primitive atmosphere may have been composed largely of CH_4 and NH_3. However, the equilibrium constants for reactions of these and other gases, combined with the evidence of the "excess" volatiles, indicate that CO_2 and N_2 are much more likely.

The stabilities of methane and ammonia depend upon the presence of free hydrogen; and the escape rate of hydrogen from the earth is such that methane probably could have persisted in significant amounts in the early atmosphere no more than 10^3 to 10^8 years. For all but a relatively brief period at the very beginning of earth history, the atmosphere probably contained CO_2 and N_2 rather than CH_4 and NH_3.

When the consequences of a dense atmosphere of CO_2 and N_2 (but with almost no free O_2 or H_2) are examined, it is found that several chemical effects (such as the quantity of rocks that would have to be weathered, of sodium dissolved in sea water, and of $CaCO_3$ deposited on the sea floor very early in early history) are not borne out by the observed geologic record.

From this and other lines of evidence it seems extremely improbable that the present atmosphere and hydrosphere are residual from any such dense primitive atmosphere. Instead, it seems likely that the atmosphere and hydrosphere have accumulated gradually during geologic time by the escape of water vapour, CO_2, CO, N_2, and other volatiles from intrusive and extrusive rocks that have risen more or less continuously from the deep interior of the earth.

The amount of free oxygen in the early atmosphere is a separate problem that cannot be solved until the evidence of the earliest rocks has been appraised more fully. Current hypotheses of the origin of life appear to require a reducing atmosphere, yet it seems likely that oxygen has been accumulating from the photodissociation of water vapour ever since the earth was formed. The oxidation of ferrous iron and sulfides in the earliest sediments may have kept the oxygen content very low, and life may have begun in local reducing environments.

In a symposium on the crust of the earth no apologies are necessary for considering the probable history of air and water, even though they are not hard rocks. If we could be sure that the earth was somehow born fully equipped with ready-made continental masses and a crustal layer, then the hydrosphere and atmosphere would probably have influenced the rocks of the crust only superficially, by way of such processes as the erosion of uplifts, the deposition of sediments, and the development of living organisms. But if, as many geologists and others think likely, the crust, the continents, and the ocean basins have evolved from changes and dynamic processes that have operated through the geologic past, then the hydrosphere and atmosphere are almost certainly inter-related in origin and subsequent history with the deeper-seated processes of petrogenesis, epeirogeny, and mountain making.

Just what the mechanism of this inter-relationship may be is to the writer a fascinating problem that touches many aspects of earth history but its discussion will not be attempted in this paper.

The origin and subsequent history of the hydrosphere and atmosphere are still unsolved problems and probably will long remain so. The questions involved in these problems depend for their answers upon evidence from a surprising range of scientific disciplines — not only the fields of geology, geochemistry, meteorology, oceanography, and hydrology, which come readily to mind, but also such seemingly unrelated subjects as astronomy, seismology, nuclear physics, biochemistry, and others.

A truly satisfactory hypothesis of the development of the hydrosphere and atmosphere must wait until nearly all other broader problems of earth his- tory and constitution have been solved. But precisely because it is so closely related to many other problems, any progress that can be made toward unravelling the history of the earth's air and waters touches also these many other problems and thereby limits, at least to some degree, the range of permissible speculation regarding them.

It would be difficult if not impossible to make specific acknowledgments to the many colleagues who in informal discussions have contributed at least indirectly many of the ideas presented in this paper. The writer is especially indebted to his friends G. P. Kuiper and H. C. Urey of the University of Chicago and G. C. Kennedy and George Tunell of the University of California at Los Angeles for reading the manuscript and offering many helpful suggestions and criticism. None of the anonymous colleagues or four reviewers, however, are to be held responsible for any of the views expressed, as one of them has kindly been at some pains to make clear.

ALTERNATIVE THEORIES OF ORIGIN OF THE HYDROSPHERE AND ATMOSPHERE

Several years ago the writer reviewed the available evidence that bears on the origin of sea water. This review sought to assemble critical evidence, to find or formulate alternative hypotheses that appear consistent with the facts, and then to examine these hypotheses for consequences that can be tested by the actual geologic record.

The present paper outlines those portions of this earlier review that bear upon the probable chemical composition of the earth's early atmosphere and seem to merit more detailed consideration at this time.

Source of "excess" volatiles

When the probable quantities of rocks weathered and sediments deposited during geologic time are compared, it seems evident that most of the major rock-forming elements (Si, Al, Fe, Ca, Mg, Na, K, etc.) in sedimentary rocks and all the dissolved bases in sea water have been derived from the weathering of earlier rocks throughout the past.

This is not, however, an adequate source for a group of other materials (H_2, CO_2, C1, N, S, and several others), all of which are much too abundant in the present atmosphere and hydrosphere and in ancient sedimentary rocks to be accounted for solely by rock weathering. For these materials (which for convenience may be called the "excess" volatiles) some other source is required. In seeking this other source, we encounter head-on the central problem of the origin of the hydrosphere and atmosphere.

Only two possible sources of these "excess" materials appear to have been suggested. Either the waters of the present ocean and all the other "excess" volatiles have been inherited from a primitive ocean and atmosphere, or they have risen to the surface from the earth's interior during the course of geologic time.

The first possibility is sometimes stated about as follows: If the earth were once molten throughout, then much of the water and many materials were at one time volatilized in a hot primitive atmosphere. Later, as the earth cooled, the water vapour condensed and formed a primitive ocean. According to this hypothesis, the present atmosphere and ocean are simply residual from a hot primitive atmosphere.

The alternative hypothesis lacks the simplicity and vividness of the first one. This explanation depends upon some complex and relatively unfamiliar process or processes of "degassing" of the rocks of the earth's interior, and these complex processes get the hypothesis deep into problems of physical chemistry and petrogenesis.

Two distinct variants of this second hypothesis may be recognized:

- That the water and other volatiles came from the earth's interior within some brief period very early in earth history;
- That these volatile substances have been escaping from the earth's interior gradually and at about the same rate throughout much of geologic time.

"Excess" volatile materials in present atmosphere and hydrosphere and in buried sedimentary rocks after correction for quantities released by rock weathering (in units of 1020 grams)

H_2O	16,600
Total C as CO_2	910
C1	300
N	42
S	22
H	10
B, Br, A, F, etc	4

Chemical Consequences of Dense Primitive Atmosphere

In the review paper previously mentioned, an effort was made to follow through, in a semiquantitative way, some of the consequences of these alternative hypotheses, and to find corollaries of one or the other that would permit critical tests by field or laboratory evidence.

One of these several lines of consequences from alternative hypotheses was an effort to estimate the chemical effects that might be expected:

- If all the "excess" volatiles were present at one time in an early ocean and atmosphere (regardless of whether inherited from a primitive atmosphere or erupted suddenly from the interior early in earth history).
- If, on the other hand, the volatiles accumulated gradually to form the present ocean and atmosphere.

With a primitive ocean and atmosphere made up of all the "excess" volatiles indicated in Table, the different elements and compounds would dispose themselves between solution in the ocean water and as gases in the atmosphere, depending on their individual chemical properties and their quantities.

Under the postulated conditions and with the given gross composition (which would permit essentially no free oxygen or hydrogen), it was assumed that, at the outset before rock weathering became important, the predominant gases would be CO_2, N_2, and H_2S, rather than CH_4, NH_3, and other possibilities. With an assumed temperature of 30°C as a basis for estimating the distribution of materials between ocean and atmosphere, it becomes possible to work out the sequence of chemical changes that would be expected.

On this hypothesis the ocean and rain water would at first be intensely acid. This acid water would decompose bare rock with which it came in contact and in so doing would dissolve some of the bases from these rocks.

As a result, the acidity of the water would decrease, and the chemical effects of dissolved gases would come into play. As decomposition of rocks proceeded, more and more dissolved bases, chiefly Ca, Mg, Na, K, and Fe, would be carried to the sea; as soon as the appropriate solubility products were exceeded, $CaCO_3$, FeS_2, and other compounds would be precipitated on the sea floor.

Under the postulated conditions the atmosphere would have been made up largely of CO_2 at the outset, but as carbonates continued to precipitate from the ocean more and more of the original CO_2 would be subtracted from the atmosphere. Eventually the concentration of this gas would be decreased to some point at which the most primitive forms of life might conceivably have existed.

The highest partial pressure of CO_2 that any living organism is known to tolerate may be selected, rather arbitrarily, as something that might have approximated the CO_2, content of an early Precambrian atmosphere. From this train of argument it becomes possible to deduce, as chemical consequences of a dense residual atmosphere, the quantities of rocks that would have to be decomposed, of bases that would have to be dissolved in sea water, and of $CaCO_3$ that would have to be deposited in order to reduce the original CO_2 of the atmosphere to a concentration at which primitive life might have begun.

The results of this calculation appear distinctly unfavorable to the hypothesis from which we started. The amount of rock that would have to be decomposed under this hypothesis before life began would be very large — considerably larger in fact than seems, from independent estimates, to have been decomposed in all of geologic time, both before and after the advent of living organisms.

Similarly the total amount of Na dissolved in sea water would have been much greater, even before life began, than the amount which from other evidence appears to have been dissolved in all earth history — that is to say, the hypothesis leaves no place for solution of any additional Na after early Precambrian time. Furthermore, the quantity of carbonate deposits required means that the early Precambrian should have been a time of abnormally great carbonate deposition throughout the world; more limestone and dolomite should have been deposited then than now remains in rocks of all ages.

The known geologic record indicates relatively few thick limestones of early Precambrian age; the record suggests instead that the bulk of the earth's limestone was deposited much later, during Palcozoic time.

Chemical Consequences of Gradual Accumulation

If the same type of reasoning is now applied to the alternative hypothesis (*i.e.*, that the "excess" volatiles have accumulated gradually to form the present ocean and atmosphere), the results are significantly different.

The same assumptions may be made regarding which gases would be stable, the decomposition of rocks, and the solution of bases by acid waters. But where for the other hypothesis an upper tolerance limit of CO_2, in the atmosphere was assumed to measure the quantity of rocks decomposed, Na dissolved, and carbonates deposited before life became possible, for this alternative hypothesis we may assume that the CO_2 content of the atmosphere never rose above this same limit.

From these assumptions calculations show that carbonates, for example, would begin to precipitate from the sea water when less than one-tenth of the total water and other "excess" volatiles had accumulated in the atmosphere-ocean system. Thereafter, if CO_2 continued to be released into the system at a relatively constant rate, rock decomposition would proceed, and limestones would be deposited at a more or less constant rate down to the present time.

That is to say, with this alternative hypothesis we are not led to expect embarrassingly large quantities of rocks decomposed, Na dissolved, and carbonates deposited in the early Precambrian, as we were with the first hypothesis.

The train of chemical consequences just outlined is but one of several independent corollaries that may be deduced from the opposed hypotheses. These will not be discussed here, except to mention that all such corollaries that the writer has considered and has tried to follow through semiquantitatively seem to lead rather compellingly to the same conclusion — namely, that the geologic evidence appears distinctly to favour the hypothesis that the atmosphere and ocean have accumulated gradually from some process of "degassing" of the earth's interior and that this process must have operated at a more or less steady rate through geologic time.

GASES IN THE EARLY ATMOSPHERE

In trying to estimate the initial chemical character of the postulated early atmosphere and ocean, the writer had followed traditional theory in taking carbon dioxide, free nitrogen, and hydrogen sulfide as the principal gases likely to be stable under the assumed conditions.

However, a number of investigators in various fields have recently considered somewhat the same general problem and have taken other gases, particularly methane or ammonia, or both, as more likely to have been the dominant constituents of the earth's early atmosphere. Poole also at one time advocated the prevalence of methane in the early atmosphere but more recently he has rejected this view and concluded instead that a primitive atmosphere of carbon dioxide, free nitrogen, and water vapour seems more probable.

The reasons that have led these writers to consider methane or ammonia, or both, as major constituents of the early atmosphere are probably several, but they may include one or more of the following considerations:

First, we know that hydrogen and helium greatly exceed in abundance all other chemical elements in the spectra of the sun and stars. For significant downward revision of previous estimates of the relative abundance of helium, *see* Neven Underhill. If hydrogen were at one time very abundant in the atmosphere of the earth, then methane and ammonia, rather than carbon dioxide and nitrogen, should have been the dominant gases.

A second consideration is the fact that methane and ammonia are the most abundant gases in the atmospheres of the major planets, Jupiter, Saturn, Uranus, and Neptune. These planets are all more distant from the sun and more massive than the earth, and it does not necessarily follow that these gases were once present in the same abundance on earth.

Third, the hypothesis of Oparin and Horowitz is widely attractive to scientists in many fields. This postulates that before ozone became a significant constituent of the earth's atmosphere, complex organic compounds were synthesized by photochemical processes; that the most primitive forms of life originated in this way; and that these first self-duplicating molecules evolved into more specialized organisms as they consumed the supply of previously formed organic compounds.

This hypothesis seems to require a reducing atmosphere, and some of its advocates have held that it also requires the presence of either methane or ammonia, or both. Others, however, seem to think that the same general principle could operate quite as well in an atmosphere of carbon dioxide, provided no free oxygen were present.

Finally, Miller has succeeded in synthesizing two amino acids, the essential substances of proteins, by passing an electric discharge (the effects of which would be comparable to lightning) through a mixture of water vapour, methane, ammonia, and hydrogen. Insofar as known, this experiment has not yet been tried with other combinations of gases.

The chemical character of the early ocean would of course have depended greatly upon what gases were dissolved in it. If the calculations in the writer's "Sea water" paper were based on the wrong gases in the atmosphere, then the specific conclusions reached regarding the probable chemical effects of a dense residual atmosphere and ocean would require modification.

It should be mentioned, however, that the over-all conclusion reached there — that the earth's atmosphere and hydrosphere are probably secondary and have accumulated gradually from "degassing" of the interior — is not affected by this question of which gases were dominant in the early atmosphere, because it rests upon several independent lines of evidence, such as the mineralogical and paleontological evidence that CO_2 must have been supplied to the atmosphere at a fairly constant rate throughout the past.

As indicated by the list of writers who have discussed various aspects of the subject recently, the question of the probable composition of the earth's early atmosphere is of considerable interest in a number of fields of science.

The evidence afforded by the "excess" volatiles appears to bear directly on this question and to place some restrictions on the range of permissible speculation regarding it. Inasmuch as this evidence of the "excess" volatiles seems not to have been considered as fully as its importance warrants, the writer has presumed to wander far outside his own field of science in order to present at least the high lights of it.

EQUILIBRIUM CONCENTRATIONS OF DIFFERENT GASES

Gross Composition of Dense Primitive Atmosphere from "Excess" Volatiles

The significance of the "excess" volatiles for this problem may be seen most readily if one tries to estimate the probable composition of an early atmosphere under the two alternative hypotheses previously described:

(a) Residual from a dense primitive atmosphere.
(b) By gradual accumulation.

The basic assumption is that the "excess" volatiles (that is, those materials present in today's atmosphere, hydrosphere, and bio- sphere and entombed in ancient sediments that cannot be accounted for by rock weathering) afford the best, and in fact almost the only, direct evidence of the probable composition of the early atmosphere, whatever may have been its mode of origin.

In order to learn which gases would be stable under various

conditions, the writer's earlier estimates of "excess" volatiles may be taken as a starting point. Perhaps these earlier estimates could now be improved, but they are at best only estimates, and these readily available figures are used for the present calculations in the belief that any conclusions based upon them will still be valid qualitatively even when these estimates are improved by new data.

Table. Chemical Reactions Considered

(g) indicates the gas phase				
H_2O (g)			=	H_2 (g) + ½O_2 (g)
CH_4 (g)	+	$2H_2O$ (g)	=	CO_2 (g) + $4H_2$ (g)
CH_4 (g)	+	H_2O (g)	=	CO (g) + $3H_2$ (g)
CH_4 (g)			=	C (solid) + $2H_2$ (g)
NH_3 (g)			=	½N_2 (g) + 3/2H_2 (g)
2HC1 (g)			=	H_2 (9) + C12 (g)
H_2S (g)			=	S (solid) + H_2 (g)

Equations were then assembled of a number of gas reactions that seem likely to be important for this gross composition and for moderate temperatures and pressures. To keep the total number of components small, the possible oxidized gases of sulfur are neglected. This neglect probably introduces no significant errors: of the four principal gas-forming elements to be considered, sulfur is the least abundant among the "excess" volatiles; furthermore, with the gross composition given, little if any free oxygen is possible, and the sulfur oxides would therefore be negligible in amount.

The mean temperature of the present atmosphere, considering the distribution of its mass and temperature with height is approximately – 20°C. The conditions of the present inquiry require an average surface temperature consistent with a liquid ocean but are not otherwise more narrowly specified.

Experimental data for oxidation potentials, free energies of formation, and equilibrium constants are conveniently tabulated for the standard laboratory temperature of 25°C. If this be taken as the mean temperature of the postulated early atmosphere, it means, by comparison with the present earth, a climate tropical even at the poles.

In view of numerous other sources of uncertainty in the calculations and estimates that follow, it seems scarcely worthwhile to attempt the thermal corrections necessary to convert the equilibrium constants to a mean temperature near that of the earth's atmosphere at the present time. For these reasons 25°C has been used as the basis for calculation.

The several degrees of dependence of these equilibrium mol

fractions upon pressure, here measured by the exponents of the symbol P, are based on the ideal gas laws. These pressure relationships would not hold strictly true, of course, for actual gases, but they serve as a useful basis for approximation.

$\frac{(H_2)\cdot(O_2)^{-1/2}}{(H_2O)}$	$= 9.0 \times 10^{-41} \times P^{-1/2}$
$\frac{(CO_2)\cdot(H_2)^4}{(CH_4)\cdot(H_2O)^2}$	$= 1.3 \times 10^{-20} \times P^{-2}$
$\frac{(CO)\cdot(H_2)^3}{(CH_4)\cdot(H_2O)^2}$	$= 1.3 \times 10^{-25} \times P^{-2}$
$\frac{(Sol)\cdot(H_2)^2}{(CH_4)}$	$= 1.3 \times 10^{-9} \times P^{-2}$
$\frac{(N_2)^{1/2}\cdot(H_2)^2}{(NH_3)}$	$= 1.2 \times 10^{-3} \times P^{-1}$
$\frac{(H_2)\cdot(Cl_2)}{(HCl)^2}$	$= 2.0 \times 10^{-17} \times P^0$
$\frac{(Sol)\cdot(H_2)}{(H_2S)}$	$= 1.6 \times 10^{-6} \times P^0$

P = Sum of partial of gaseous components participatng in given reaction.

With the seven equilibrium constants and the six totals of C, C1, N, S, O, and H from the estimates of "excess" volatiles we have 13 equations and 13 components of unknown concentration participating in the seven reactions. Theoretically at least, this problem can be solved. In fact, however, the writer found it intractable. The number of unknowns was therefore reduced to 12 by assuming that an early atmosphere in contact with an extensive ocean would be saturated with water vapour.

For the assumed temperature of 25°C and the total pressure of all gases (which is to be found later in the calculation), this gives the number of mols of water vapour in the atmosphere. The scarcely perceptible effects of this arbitrary assumption upon the numerical values of the final results are mentioned more explicitly below.

In preparing the list of "excess" volatiles given in Table, one finds that some of the chemical elements are more readily tallied in such an inventory than others. Carbon and chlorine, for example, are readily enough recognized as such wherever they occur, but the "excess" oxygen and "excess" hydrogen are relatively subtle concepts, and these materials are thus more difficult to identify with reasonable assurance.

Strictly speaking, it is not quite logical to accept the estimates of both "excess" hydrogen and "excess" oxygen, for it would be simpler and possibly more accurate to assume that all the "excess" hydrogen was originally combined as H_2O, thus leaving no "excess" hydrogen. But to make this assumption would prejudice the outcome of the present inquiry as it would mean almost no possibility of any CH_4 or NH_3 in the early atmosphere.

It was therefore decided to use for these calculations all the 10×10^{20} g of estimated "excess" hydrogen. As the number of unknowns had, for reasons mentioned above, been reduced from 13 to 12, the number of equations required for solution of the problem was also reduced to 12; because of the uncertainties involved in it, the separate estimate of 506×10^{20} g of "excess" oxygen was not used in these calculations.

One still needs to assemble the experimentally determined solubilities of each of these gases in water. When these solubility data and the equations are all duly tabulated, only the work of solving the problem remains. This proved to be a formidable chore.

The amount of a gas dissolved depends upon the quantity of water, the solubility of the gas, and its partial pressure in the atmosphere. But the partial pressure of the gas depends in turn upon how much of the total amount is dissolved in water and how the amount that still remains in the atmosphere compares with the total of all other gases present. And so on, almost *ad infinitum*. However, with sufficient patience, the problem can be solved numerically by a series of successive approximations.

When all the paper work is done, we have an estimate of the amounts of the different gases under equilibrium conditions. It is, of course, only an estimate. Perfect equilibrium might never be attained, even with geologic time. Ammonia would react with CO_2 and HC1 to form carbonates, chlorides, and other compounds; these would dissolve in water but scarcely affect the distribution of the gases between atmosphere and ocean. Furthermore, no account has been taken of photochemical processes that would have operated on the gases.

Such photochemical reactions would doubtless form other and more complex compounds. But from what is known regarding the photochemical destruction of ammonia and methane, the net effect of such processes would be to decrease rather than to increase the proportion of these two gases. Insofar as the results have significance then, they mean that, under the assumed conditions, free nitrogen in the atmosphere would at the outset tend to be about 490 times more abundant than ammonia, and carbon dioxide about 3800 times more abundant than methane.

The arbitrary assumption of a water-saturated atmosphere made to facilitate the computations scarcely affects the numerical value of the final results. Table shows that, even if fully saturated with water vapour.

Table: Composition of Early Atmosphere and Ocean by Residual Hypothesis

	Atmosphere (per cent by volume)	Dissolved in ocean (units of 1020 gm)
CO_2	91	240
N_2	6.4	0.2
H_2S	2.0	12
HC1	0.2	310
H_2O (vapor)	0.2	
NH_3	0.01	14
H_3BO_3		7.4
HBr		1.0
HF		0.3
CH_4	0.02	0.004
Cl_2	0.01	0.2
CO	*	†
H_2	*	†
	Total atmospheric pressure 14.1 kg/cm²	Mass of ocean water -- 16.6×10^{20} kg pH of ocean water - - 0.3
* Less than 1 ppm.		
† Less than 1 ppm of total dissolved substances.		

The total atmosphere required by this hypothesis is so large that water vapour can make up only a very small part of it. For this reason the effects of this arbitrary assumption upon the final results are negligible. To appraise these effects, one may assume an atmosphere only one-tenth (instead of fully) saturated with water vapour and then carry through the same calculations as before.

When this is done it is found that, of the volume percentages listed in the centre column of Table, only that of water vapour itself is modified by this change: it becomes 0.02 instead of 0.2 per cent; the others are changed by amounts too small.

It is important to note that the composition indicated in Table 4 is that of the postulated early atmosphere while still uncontaminated by the effects of rock weathering. As mentioned above, this intensely acid water of the ocean (and also of rain from an atmosphere of this composition) would promptly begin to decompose bare rock with which it came in contact. The bases thus dissolved from silicate minerals would accumulate in the sea until they reached concentrations sufficient to cause precipitation of $CaCO_3$, FeS_2, and other compounds.

Such precipitation would subtract acid radicals from the sea water, and consequently the composition of the atmosphere would in time be modified significantly. With an atmosphere of this original composition the most notable effect would be the subtraction of large quantities of CO_2.

The results of these calculations thus seem to confirm what was simply assumed in the previous paper — namely, that, with a gross composition equivalent to that of all the "excess" volatiles, the postulated dense primitive atmosphere would at the outset have been made up dominantly of carbon dioxide and nitrogen.

Gross Composition from Gradual Accumulation of "Excess" Volatiles

It is scarcely necessary to carry through the laborious calculations still another time in order to see which gases would be the dominant ones under the alternative hypothesis of gradual escape of the "excess" volatiles from the earth's interior. Under this hypothesis the total atmospheric pressure would never have been great at any time. Neglecting for the moment the effects of rock weathering, the principal differences in relative abundances of the several gases in the atmosphere would be those due to the smaller volume of ocean water and to the effects of lower partial pressures of the different gases.

Under these conditions smaller quantities of all gases would be dissolved in ocean water, and relatively more would therefore remain in the atmosphere. For example, carbon dioxide, which is more soluble than methane, would be relatively more abundant in the atmosphere under this gradual-accumulation hypothesis than under the alternative one of higher partial pressures of the gases and greater volume of ocean water.

A second and quantitatively more important difference depends upon the fact that the equilibria of those gas reactions that are pressure dependent would, under the gradual-accumulation hypothesis, be shifted toward the components that occupy more volume. If, for example, the total atmospheric pressure were 1 kg/cm^2 instead of 14.1 kg/cm^2, the equilibrium ratio of CO_2 to CH_4 would be $(14.1)^2 = 200$ times greater than under the first hypothesis, other things being equal.

Thus it seems clear that, under the hypothesis of gradual escape of volatiles from the interior, even more than under the other hypothesis of a dense primitive atmosphere, CO_2 and N_2 would be the dominant gases to be expected initially. As volatiles continued to leak upward gradually from the "degassing" of the earth's interior, they would become mixed with an atmosphere and ocean already somewhat modified in composition by the effects of earlier rock weathering and deposition of carbonates.

With gradual addition of more volatiles and especially if free oxygen began to accumulate from the photodissociation of water vapour in the upper atmosphere and from the photosynthesis of carbon from CO_2 by plants, the compositions of both atmosphere and ocean water would approach ever more closely those of today.

OTHER LINES OF EVIDENCE

The present inquiry might be terminated at this point with the conclusion that the hypothesis of gradual "degassing" of the earth's interior leads to chemical consequences at the surface that appear entirely consistent with the observed geologic record. But to close the discussion here would be to neglect other considerations and lines of evidence that deserve attention.

It is obvious that some gross composition of the atmosphere and hydrosphere very different from that of the "excess" volatiles would lead by this method of calculation to a very different composition of the early atmosphere. For example, one might assume, entirely *ad hoc,* an original source of methane and ammonia and thereby reach entirely different conclusions about the atmospheric composition.

However, without independent evidence to support it, such an assumption would seem unwarranted. A brief survey is therefore attempted to see if lines of evidence neglected in the above discussion may point to some gross composition very different from that of the "excess" volatiles.

If, as seems probable, the atmosphere has grown by "degassing" of the earth's interior, one might look for evidence of some different gross composition of the early atmosphere in the gases now escaping from volcanoes and hot springs or in the gases occluded in igneous rocks and meteorites. Reliable data on the composition of such volcanic and occluded gases are scanty. In most gases from volcanoes and hot springs and in gases occluded in igneous rocks that have been analyzed, CO_2, CO, and N_2 greatly predominate over CH_4 and NH_3, and the average composition of such gases agrees reasonably well with the gross composition of the "excess" volatiles.

This evidence is not so conclusive as it sounds, however, because it is uncertain to what extent such gases may have been contaminated by previous contact with the atmosphere or with sedimentary rocks and then recycled through a later stage of molten rock. Because of this possibility of contamination and the common view that meteorites probably afford the best available clues about the composition of the earth's interior, the occluded gas content of meteorites may be particularly significant, for meteorites presumably have not been subjected to previous contact with the earth's atmosphere or to subsequent recycling.

The scanty data available indicate that the occluded gases in stony and iron meteorites are likewise relatively high in, CO_2, CO, and N_2 and low in CH_4 and NH_3. Admittedly the analysis of occluded gases is difficult and the results uncertain, because the combination of chemical elements can be modified significantly by the analytic extraction procedures used. But even with these uncertainties in mind, no important sources of methane and ammonia are indicated.

In the gases of stony meteorites, the type presumably most nearly representative of the silicate rocks of the deep interior, the atomic volumes of oxygen exceed the combined totals of carbon, hydrogen, and nitrogen, and oxygen exceeds hydrogen by an average factor of about four. The available evidence therefore suggests that the gases now escaping from the earth's interior and also the occluded gases as an index of those most likely to escape, are of approximately the composition of the "excess" volatiles and that CH_4 and NH_3 are relatively uncommon.

If no reason seems evident for assuming an important source of CH_4 and NH_3 in the gases from the earth's interior, one may look for some process that might have produced these gases at the earth's surface or within the atmosphere. Ammonia has not been found, but small quantities of methane and hydrogen have been detected in the present atmosphere.

Both these gases are subject to known dissipative processes. Methane undergoes decomposition by normal oxidation and by photochemical, biological, and other processes and hydrogen is so light an element that it is escaping from the earth's gravitational field. It seems reasonable to assume that both gases are in a steady state and are being continually produced by processes that operate today and thus may possibly have operated more actively in the past.

The small amount of methane in the present atmosphere seems fully accounted for by strictly biological processes and it therefore affords no evidence of an important generative process that might have operated before the advent of life.

The hydrogen in today's atmosphere is of interest in the present inquiry because, if the amount were much greater sometime in the past, the chemical equilibria are such that CH_4 and NH_3 would also have been relatively abundant. According to Paneth the atmosphere now contains about 5×10^{-7} parts by volume of H_2 or approximately 1.8×10^{14} g in the total atmosphere. This amount may be compared with the yield of the only natural process that is thought to produce hydrogen in significant quantities today.

In the upper limits of the atmosphere, H_2O vapour is probably undergoing continual decomposition by ultraviolet radiation. After reviewing the pertinent data, Kuiper and greatly lowered previous estimates of the rate at which this process operates. He concluded that approximately 7.7×1.0^{-16} mol $H_2O/cm.^2$ see are being decomposed, which would mean the production of 2.5×10^{11} g H_2/yr.

From Spitzer's modification of Jeans' fundamental equation for the escape rates of lighter atoms and gases from a planet, one may estimate that this quantity of hydrogen would escape from the 1.8×10^{14} g of the present atmosphere each year if the kinetic temperature of the upper atmosphere were approximately 3500°K.

However, the temperatures thought to prevail in the upper atmosphere today, 500 km or so above the earth's surface, are around 1500°K, with occasional bursts of radiation producing a maximum of something like 2500°K. Taken at face value, therefore, this calculation suggests that hydrogen is now being produced more rapidly than it is escaping — in other words that it is accumulating toward some concentration at which the escape rate can eventually equal the production.

It is much more probable, however, that the discrepancy between the calculated and actual temperatures of the upper atmosphere is attributable solely to uncertainties in the data used for this calculation. If the rate of photodissociation is less than Kuiper estimates or if the hydrogen content of the uppermost atmosphere is greater than Paneth's estimate, the discrepancy disappears.

The results of this calculation are merely suggestive of course. Spitzer points out that his equation is not reliable for very rapid escape rates, because with such rates diffusion of the gas becomes the controlling factor. However, the available information indicates that the present rate of photodissociation of water vapour in the upper atmosphere is probably quite adequate to explain the small amount of H_2 observed in today's atmosphere.

If for any reason the rate of photodissociation of water vapour were much greater sometime in the past while the escape rate of hydrogen remained the same, or if the upper atmospheric temperature which controls escape rates were much lower while the photodissociation rate remained the same, then the concentration of hydrogen in the atmosphere might have been greater than it is today.

Presumably, however, the rates of the two processes vary together with any changes in upper atmospheric temperature, and, without convincing reasons why some difference in the balance between the two processes should be expected in the past, it is unwarranted to assume that photodissociation of water vapour has ever caused a significantly higher concentration of hydrogen in the atmosphere than it does today.

The gases escaping from the earth's interior and those being produced at the surface afford no strong evidence of important sources of CH_4 and NH_3 in the past. But when we turn to consider possible sources from out in interplanetary or interstellar space, a gross composition very different from that of the "excess" volatiles is clearly suggested. The question then becomes whether such a possible atmosphere from outer space could have persisted long enough after the earth was first formed to leave chemical evidence of itself in the earliest rocks and in the subsequent history of the earth.

The hypothesis that the present atmosphere and ocean are largely residual from a dense primitive atmosphere fits particularly well with the possibility that the earth may have brought with its solid substances from outer space an initial atmosphere very different in composition from that calculated above.

This supposed initial atmosphere might have consisted not only of the "excess" volatiles as postulated above but also of all other chemical elements in the proportions observed in the solar and stellar atmospheres today. This would mean a very high concentration of hydrogen, which in turn would greatly favour the occurrence of CH_4 and NH_3 rather than CO_2 and N_2.

To make this assumption explicit, one might start with a primeval earth containing the present percentages of Si, Fe, and other heavier elements and calculate from cosmic abundances the quantities of lighter elements it should have contained originally. This would give a protoearth approximately 190 times as heavy as the present earth and consisting of 80 to 85 per cent by weight of hydrogen.

After allowing for oxygen to balance Si, Mg, etc., in the rock forming oxides, that remaining affords a measure of the huge volume of water to be expected on such a planet. To this set of assumed conditions the various gas-reaction constants and solubilities in sea water may then be applied as in the calculation described above for the "excess" volatiles. This yields an initial atmosphere of about two-thirds helium and one-third hydrogen by volume, and minor amounts of other gases, among which CH_4 and NH_3 would be much more abundant than CO_2 and N_2.

Whether or not this exceedingly hypothetical set of conditions could have persisted long enough to have chemical effects still perceptible in the earliest rocks of the earth depends largely on the rate at which the lighter gases and especially hydrogen would escape from the planet's gravitational field. If the kinetic temperature at the upper limits of this initial atmosphere could be estimated, the escape rates might be calculated from Spitzer's equation.

Assuming a temperature of 1500°K, as today, the initially tremendous quantities of hydrogen would decrease to present atmospheric concentration in only 3×10^5 years. However, the upper atmospheric temperature today depends largely on the properties of the two dominant gases, N_2 and O_2, and the temperature in an He – H_2 atmosphere would be different and difficult to estimate. It would probably be as high as or even higher than that of the present $N_2 - O_2$ atmosphere, and if higher the escape rate would be more rapid.

By an argument quite independent of the temperature or even of the exact mechanism of escape — *i.e.*, by comparing the energy needed to drive off the lighter elements with that received from the sun — Kuiper estimates that a period of about 108 years would be required in the early stages of earth formation for this escape. Or if this initial atmosphere escaped before the earth condensed to its present size.

The details of the above calculations based on a protoearth composed largely of hydrogen and helium are not given here because, when one looks at this model critically, it is quite incredible. Modern views of the origin of the solar system and earth are that the great bulk of the lighter elements, and perhaps much of the heavier elements as well, escaped long before the earth condensed to anything approaching its present dimensions.

The results of these calculations are mentioned here merely as something in the nature of a limiting case. The admittedly incredible model, being very high in hydrogen at the outset, calls for an extremely high escape rate at the beginning; nevertheless, it is more favorable to the presence of large amounts of CH_4 and NH_3 in an initial atmosphere than most other assumptions one might make.

With lapse of time and depletion of hydrogen, the escape rate would become slower; and this extreme assumption then becomes approximately equivalent to other far more moderate assumptions that might be made about the original atmospheric composition.

Among recent discussions of the composition of the early atmosphere, two contributions by H. C. Urey are especially prominent.

He concludes that methane and ammonia have probably been important constituents of the atmosphere for a large part of geologic time. Many of the arguments he presents in support of this conclusion have been discussed in the paragraphs above. However some of his points seem less than convincing to the writer, and these deserve separate consideration.

In discussing the high temperatures to be expected as stony and iron meteorites accumulated to form the earth, he lists several chemical reactions that should take place under these conditions and gives the equilibrium constants of these reactions. He then remarks: "From these equilibrium constants one sees that hydrogen was a prominent constituent of the primitive atmosphere and hence that methane was as well".

It is perfectly valid procedure to deduce quantitative corollaries from an assumption and then to see whether or not these corollaries are confirmed by observed facts; but here, unless the writer misinterprets the passage, Urey seems to accept the deduced corollaries as in themselves facts. Urey merits a salutation from geologists for the equation,

$$CaSiO_3 + CO_2 = CaCO_3 + SiO_2$$

which he uses to show how limestone is formed from rock silicates and why the concentration of CO_2 in the atmosphere must have remained very low at all times. This equation, literally wollastonite + carbon dioxide = calcite + quartz, is so condensed and dramatic a metaphor that it is scarcely recognizable to geologists as a shorthand statement of the familiar over-all process by which rocks are decomposed and limestone deposits formed.

It is nonetheless a useful thumb-nail summary of the long and complex series of genetic relationships usually spelled out, laboriously but in a still oversimplified version, somewhat as follows:

- Carbon dioxide, sunlight, and water favour the growth of plants on the land and in the sea.
- Plants support the growth of all other organisms.
- Decaying organic matter yields CO_2 which, dissolved in ground water, forms carbonic acid.
- Calcic feldspars, pyroxenes, and amphiboles in crystalline rocks and carbonates in sedimentary rocks are decomposed by weak solutions of carbonic acid, with the formation of clay minerals and the release of Ca and other bases into aqueous solution.
- Ground waters carrying Ca and other dissolved materials discharge into streams and so into the ocean where solar

evaporation causes gradual concentration.

- CO_2 in the atmosphere produces a small but significant concentration of CO_3 — ions in sea water.
- When the solubility product of CA^{++} ions and CO_3-ions exceeds a certain value, $CaCO_3$ is deposited.

It so happens that an *increase* in dissolved CO_2 *decreases* the concentration of dissolved CO_3 — ions rather than the reverse so that an increase of atmospheric CO_2 would cause solution rather than precipitation of $CaCO_2$, thereby allowing Ca^{++} ions to accumulate in the sulfate-bearing sea water until gypsum as well as calcite would become insoluble.

Consequently, at first glance, Urey's chemical shorthand looks exactly backwards for the deposition of limestone. But looked at more carefully it is seen to be entirely valid, because in the long run the formation of carbonate deposits in the sea depends upon the supply of calcium from rock weathering on the lands; and rock weathering in turn depends, by way of decaying organic matter, upon the concentration of carbon dioxide in the atmosphere.

Properly interpreted, Urey's equation is a brilliant and useful generalization regarding the net effect of two basic and highly complex processes in the chemistry of the earth. Read literally, however, it certainly does not accurately represent the way limestone deposits are formed in nature.

Even the broader relationship it expresses so succinctly between atmospheric CO_2 and rock weathering must be viewed with circumspection, because silicate rocks decompose very slowly in pure rain water that is uncontaminated by decaying organic matter. In the writer's opinion, this equation cannot properly be used to estimate even approximately the maximum concentration of CO_2 in the early atmosphere before life had appeared on earth.

This brief survey of the evidence from volcanic and occluded gases, from present-day processes by which CH_4 and H_2 are being formed, and from the cosmic abundances of elements has turned up no positive evidence of a gross composition of the atmosphere and hydrosphere that was significantly different any time in the past from that indicated by the "excess" volatiles.

Calculations indicate that, with this gross composition, CO_2 and N_2 were probably the dominant gases in the earth's early atmosphere. No positive evidence is known, yet the observed facts do not contradict the possibility that the earth may have had, for a relatively brief period after it was first formed, a very different initial atmosphere which was soon lost or greatly modified by escape of the lighter elements.

The very fact that hydrogen is escaping from the upper atmosphere today precludes us from saying that hydrogen may not have been, even for a considerable time, much more abundant than it is today; and having escaped, it has left no record of itself in the "excess" volatiles that remain. If hydrogen were once abundant, then so also must have been methane and ammonia. This view is held by some writers, but from the available evidence the writer considers this possibility exceedingly remote.

PROBLEM OF FREE OXYGEN IN THE EARLY ATMOSPHERE

Little mention has been made thus far of the presence or absence of free oxygen in the earliest atmospheres. This is in itself a difficult problem distinct from those discussed above, but a few remarks regarding it seem appropriate here. The gases escaping from volcanoes, fumaroles, and hot springs and those occluded in igneous rocks and meteorites contain no free oxygen, except small amounts that seem clearly the result of contamination from other sources.

Likewise the estimates of "excess" volatile materials in the present atmosphere and hydrosphere and in ancient sedimentary rocks afford no evidence of any original free oxygen; in fact, they indicate less oxygen than enough to balance the "excess" carbon as CO_2. Thus whether the earliest atmosphere was residual from a dense primitive atmosphere or accumulated gradually by escape from the interior, no reason is known to expect free oxygen from such sources when the earth was first formed.

Just how soon thereafter oxygen began to accumulate is a problem about which it seems advisable to withhold judgment until the evidence from the earliest rocks has been appraised more extensively. The presence of somewhat higher ratios of ferric to ferrous iron in Precambrian slates than in average igneous rocks; of sedimentary iron oxides in the Precambrian iron formations of this and other continents; and of anhydrite and gypsum interbedded with limestone in the Grenville series suggests that at least some oxygen must have been available fairly early in geologic time.

Biologists have long been interested in whether free oxygen appeared before or after the advent of living organisms. The oxygen in today's atmosphere is being formed continuously by its release from CO_2 and H_2O during photosynthetic fixation of carbon by green plants. Green plants require free oxygen in order to survive in the dark, but certain primitive types of bacteria are capable of fixing carbon in the presence of hydrogen sulfide, hydrogen, or other compounds. Thus living organisms may have come into existence long before free oxygen became available, a conclusion that is consistent with the apparent requirements of the Oparin-Horowitz theory of the origin of life.

On the other hand, free oxygen is apparently being produced continuously in the upper atmosphere by the photodissociation of water vapour and the resultant escape of hydrogen from the earth's field. From the best evidence available this process seems to be operating at a rate which, continued throughout geologic time, would yield about 5 times as much oxygen as that in today's atmosphere. However, this quantity is only about one third to one sixth the total oxygen that has been removed from circulation by the oxidation of materials now buried in sedimentary rocks.

We are thus led to a paradox: Life probably could not have originated in the presence of free oxygen, yet free oxygen has probably been in process of formation in the upper atmosphere since the beginning of earth history. Possibly this dilemma can be resolved by considering the quantities of ferrous iron and sulfides that evidently were oxidized to ferric iron and to sulfate during the earliest stages of rock weathering and also the quantities of CO that probably were oxidized to CO_2 following the earliest stages of vulcanism.

Free oxygen could not have accumulated in the atmosphere in significant concentrations until this early oxidation was relatively complete. The appearance of even very small amounts of free oxygen in the early atmosphere would have had a tremendous effect on the equilibrium relations of some of the gases. For example, the reaction,

$$CH_4(g) + 2O_2(g) = CO_2(g) + 2H_2O(g)$$

at 25°C has the equilibrium constant,

$$\frac{(CO_2).(H_2O)^2}{(CH_4).(O_2)^2} = 1.9 \times 10^{140}.$$

This in effect says that methane is thermodynamically unstable in the presence of even faint traces of free oxygen, and that, if any methane were present in such an early atmosphere, it would burn to carbon dioxide. Ammonia is similarly unstable in the presence of free oxygen and oxidizes to free nitrogen.

For these reasons it seems doubtful that the iron of the Precambrian iron formations could have been oxidized and the sulfates of the Grenville series could have been formed unless O_2 was sufficiently abundant in the atmosphere to oxidize any CH_4 and NH_3 that may have been present. This need not have required more than a fraction of a per cent of O_2 in the atmosphere.

It is questionable, however, that even such small amounts of O_2 would permit development of primitive life freely exposed to the open atmosphere, as apparently is required by the current version of the

Oparin Horowitz theory. The problems here involved are exceedingly difficult, but they are truly basic to many sciences and for this reason call for much more attention than they have thus far received, either in the laboratory or in the field.

Specifically, the probable percentage concentration of free oxygen in the early atmosphere cannot be estimated even approximately until the evidence of the earliest Precambrian rocks has been examined much more extensively with this particular question in mind. From such evidence as is presently available, however, it seems to the writer that the atmosphere, even very early in earth history, must have contained at least a small amount of free oxygen; that there were then, just as there are today, local environments in the ocean and in mud pools where free oxygen was absent; and that it was in such local reducing environments that the first organisms may have come into existence.

Chapter 11

Folds and Faults

There is a curious kind of sandstone that can easily be bent by hand when it is in thin slabs. We say it is "curious" because everyday experience tells us that rocks are quite brittle. They may be chipped with a hammer, and one would be surprised to find a rock that merely bends under the blows. Our piece of sandstone, however, is a sort of trick rock. It is composed of quartz grains rather loosely held together with flakes of another mineral (mica); and as the rock is bent, the sand grains readily slide over the mica flakes.

Despite the brittleness of common rocks, we do find them in nature contorted into all kinds of intricate folds and plications. This is especially true in mountainous regions, where diastrophic forces have acted on the rocks. As a matter of fact, the conditions of pressure, temperature, composition, and the time element itself, are the factors that largely determine whether rocks will bend or break.

Our direct experiences with rocks occur in the surface environment with its low pressures and relatively low temperatures. It is under such circumstances that rocks break rather than fold. Further, the time at our disposal is limited. Even so, a century or two is sufficient for some changes. For example, covers of old tombs occasionally are found to sag downward a bit. The rock (commonly limestone) adjusts itself to its unsupported condition by a slow bending, even under surface conditions.

The pressure and temperature effects may be brought out by experiments that have actually been performed in laboratories. Certain substances, like salt, ice, and calcite—even some types of rock—can be made to flow when they are confined by great pressures. By *flow* we mean a fractureless change of form, for under great confining pressures the material cannot break or shatter, but must adjust itself to its environment by an actual plastic yielding.

The flow of rocks within deep-seated environments is likewise a plastic yielding, and it takes place partly through a recrystallization of the constituent minerals. The present we shall examine their larger features, the folds and other structures shown by rocks that have been subjected to diastrophic stresses in the deep-seated environment.

Suppose we consider a cubic foot of rock somewhere down beneath the earth's surface say 10 or 15 mi., in what we have called the "deep-seated environment." Our cube is thus part of a

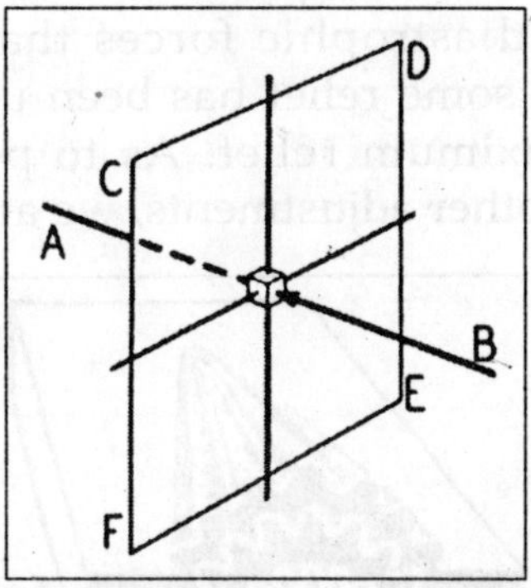

Fig. Deeply Buried Cube of Rock Acted on by Diastrophic Forces

Column of rock 1 ft. square that extends down from the surface to the depths far below. Hence it will be under considerable pressure from the overlying rocks, and at a respectable temperature. The pressure under normal conditions will be static, and the cube will remain at rest because all the forces operating on it are equal. But now suppose that, owing to adjustments in the earth, the cube is acted upon by diastrophic forces from two opposite sides. What will happen? The rock cannot very well shrink into a smaller volume, because it is already quite compact.

There may be a slight adjustment of volume due to a recrystallization of its minerals; but on the whole, the tendency will be somehow to ease the pressures that operate on it by escaping from them. The directions in which this relief will find expression lie in a plane at right angles to the diastrophic forces. In the adjacent figure the cube is acted on by diastrophic forces along the line *AB*, and the plane *CDEF* contains the directions of maximum relief.

But will the relief be upward, downward, or sideward away from the compressive force? In actual mountain-making, of course, there is a *zone* subjected to these forces, rather than some fixed cube. Hence, whatever yielding takes place affects the entire zone. Our cube, however, serves to fix attention on a particular part of the zone affected.

The question of the direction of relief cannot be answered positively, because we cannot get down to examine the situation. All we can do is to examine the eroded belts of folded rocks along mountain chains and see what they tell us. This much we know: The belts of folded rocks are usually linear, and in many cases they are symmetrical on both sides of the belt. Further, they show an elevation relative to their surroundings, so that in the folding the whole series of rocks has been uplifted.

The accompanying block diagram sketches the usual relation of a mountain belt to the diastrophic forces that formed it. We are reasonably certain that some relief has been upward and sideward along the plane of maximum relief. As to possible downfolding simultaneous with the other adjustments, we are not certain.

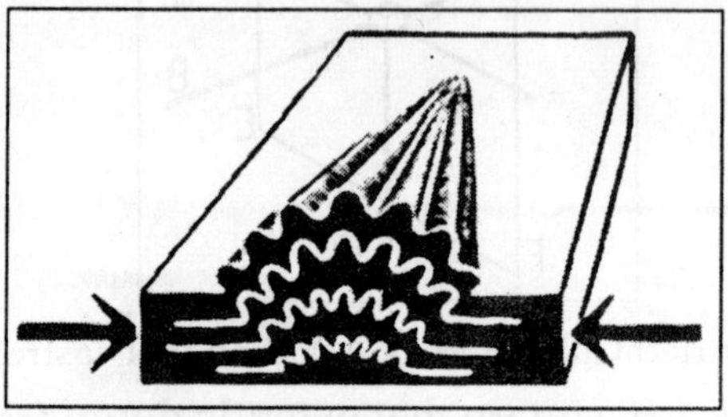

Fig. Portion of an Elevated Belt of Folds

We have to pick our way slowly and painfully, distinguishing between what we know and what we infer, for, in dealing with this subject we are approaching one of the outstanding problems of present-day geology. The nature and operations of diastrophic forces are not fully known, partly because the scale on which earth operations take place is too great for us to encompass in experiments; nor can we make direct observations at the localities where the deep-seated disturbances take place.

Confined, as we are, to surface environments, our observations include mainly the surficial aspects of these deep-seated diastrophic adjustments. The upward relief of rocks under differential stresses results in those great elevated belts of rock that we call mountains; and usually within them the rocks are found to be highly folded, or greatly mashed, and quite commonly altered from their original appearance.

Fortunately, gradational agents attack these elevated tracts and incise deep valleys among the mountain folds, so that the very roots of the mountains may be laid bare. Then, by observing how many layers of rocks are involved in the folding, and their thicknesses, we are able to tell how deeply such rocks were buried before the folding occurred.

In fact, it is by examining eroded rock structures like this that we learn most of what we know about deep-seated environments. Although we are able thus to reconstruct many of the environmental factors involved in deep-seated diastrophic adjustments, we are handicapped by the fact that the show is over.

The adjustments have been made, and we are forced to reconstruct the entire picture from the remnants we find. Fortunately again, rock-weathering is relatively slow, so that many deep-seated rocks, exposed at the surface by recent erosion, still retain most of their deep-seated characteristics.

In some cases we find that the rocks have been folded into tremendous arches, thousands of feet high, and equally broad. Such upfolds we call anticlines, and they may be rounded, domelike affairs or elongate bulges of rock.

Corresponding downfolds are called synclines. In some cases we find an alternation of upfolds and downfolds occupying a wide belt. The Appalachian Mountains are a case in point.

Fig. Anticlines and Synclines in the Jura Mountains.

Now these folds of rock need not be symmetrical or even upright. Nature is complicated in her ways, and we find folds with all degrees of complexity.

That is, if we imagine a plane through the centre and along the length of a fold (the *axial plane*) the plane will be vertical if the fold is upright; but it may be inclined at any angle, so that the folds may be inclined or even recumbent if they lie over on their side. Similarly the *axis* of the fold, which is the line along which the axial plane cuts the top or bottom of a folded stratum, may not be horizontal but may dip away at an angle.

In that case the fold is *plunging*; and by the time several of these complexities are put into the picture, the reader may be left far behind.

While an uninterested reader may dismiss the topic as abstract, the geologist dares not. On just such complexities may depend the location of valuable ore bodies or extensive pools of petroleum. For the geologist, in order to understand and interpret nature, must devise ways and means of untangling her complexities and making them available for economic exploitation or for scientific study. But these complexities of the folds are only part of the story.

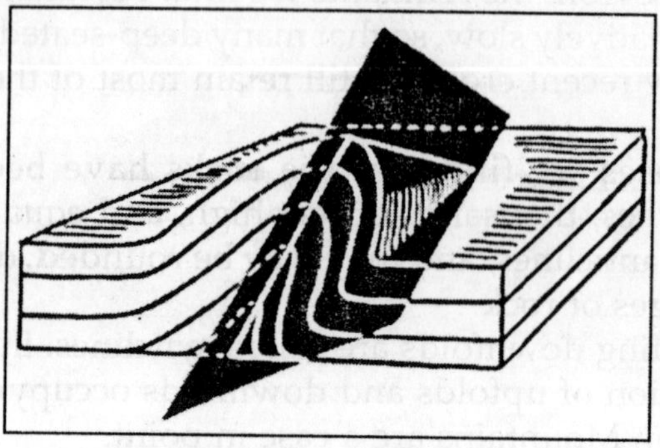

Fig. Complexities of Rock Folds

We said that erosion promptly attacks the uplifted and folded belts, so that the picture is further obscured by the absence of much of the structure and the burial of other parts of it beneath mantle rock and soil. Fortunately, the remnants of folds afford ample evidence of their former extent, and usually the bedrock is sufficiently exposed at the surface so that the essential features of the picture may be reconstructed.

When a geologist studies an area, he hunts around for outcrops, or patches of the bedrock exposed at the surface. A common place to find them is along valleys where streams have cut away the superincumbent layer of mantle rock. Ditches, railroad cuts, and other man-made excavations may also expose the bedrock. Hence you may find geologists in the most out-of-the-way places.

Several observations are usually made on the rock outcrops. For our immediate purposes we are interested in the rock structures; and to work them out, two rather precise measurements are made on each outcrop. For simplicity let us assume that we are dealing with sedimentary rocks, because then we may speak about the *attitude* of the beds. Sedimentary rocks are best adapted to such studies because they are originally laid almost horizontally, and any subsequent tilting may easily be seen. The first measurement is made along the plane of the bed, and the angular inclination of this plane from the horizontal is called the dip.

At right angles to the direction of dip, where the horizontal plane intersects the bedding, is the strike of the bed. The strike is expressed as the compass direction of this horizontal line. The adjacent diagram will give you the complete picture. Thus, dip and strike afford the necessary data to tell us what the direction and amount of inclination of the bed is at that point. In the meantime the exact location of the outcrop is also noted on a map.

Fig. Sedimentary Rocks Showing Direction of Dip and Strike

After the dips and strikes of all the outcrops have been measured and plotted, some conspicuous rock layer is chosen as a standard. Then, by connecting the indicated strikes of this rock layer, the structure of the region is determined. It may be a single syncline or anticline, miles in extent. On the other hand, there may be a succession of small folds or any combination of large and small.

Furthermore, the surface expression of the land may not reflect the structure at all, because erosional agents, cutting away in terms of the relative resistances, attitudes, and elevations of the rocks, may so disguise the original landscape that hills exist over the synclines, and valleys over the anticlines.

While plastic deformation, and its attendant rock-folding, is taking place within the deep-seated environments, diastrophic adjustments may also be taking place near the surface. Here the confining pressures are less, and the relative nearness to the surface may permit sudden crushing effects to operate on the rocks.

Hence they may yield by fracturing instead of by flow, so that joints and fractures may be formed in the rocks. If there is any slippage along the fractures, as a result of the rock on one side moving relative to that on the other, the result is a fault. Faultplanes, then, are surfaces along which relative movements take place.

When the compressive forces are unusually great, or when the rocks are unusually resistant to folding, nearly horizontal fractures may develop; and along these fractures whole mountain masses may be shoved over the underlying rocks. We call these features thrust faults, and in our own Northern Rockies such faults occurred, in one of which a great slice of rock, 10,000 ft. thick, was shoved eastward for at least 18 mi. over the rocks beneath.

This thrust fault resulted in a shortening of the earth's crust by some 18 mi. Thrust faults, then, like mountain-folding, move hand in hand with the concept of a shrinking earth.

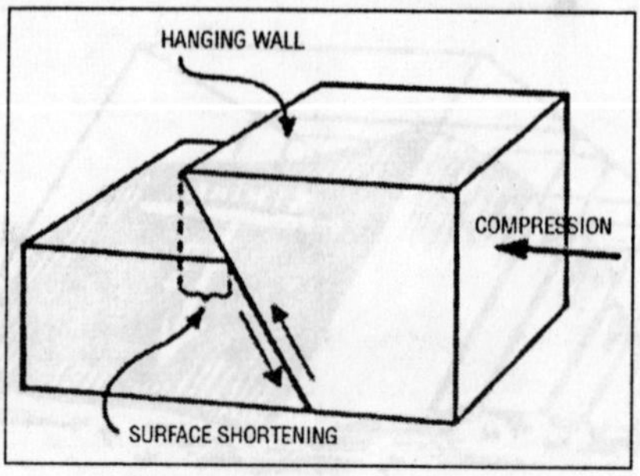

Fig. Block Diagram of a Thrust Fault, Showing Surface Shortening

In addition to thrust faults, in which the main result is a shortening of the surface, there are normal faults, which result in surface elongation. In order that you may clearly see why shortening or elongation of the surface accompanies faults, we have included the adjacent figures. Note that in both cases the fault plane dips down at an angle, so that the two surfaces which bound it may be called the hanging wall and the foot wall.

Notice that 'n the thrust fault the hanging wall has gone *up* relative to the foot wall, while in the normal fault the hanging wall has gone *down*. The resultant shortening or elongation is indicated in the figures. In actual practice it is this relation between the walls that is observed by the geologist; and from it, when he finds the evidence that shows the relation clearly, he is able to tell which kind of fault is involved and in which direction to continue the search for a faulted ore body or coal seam. Examples of faults are included in Plate 23.

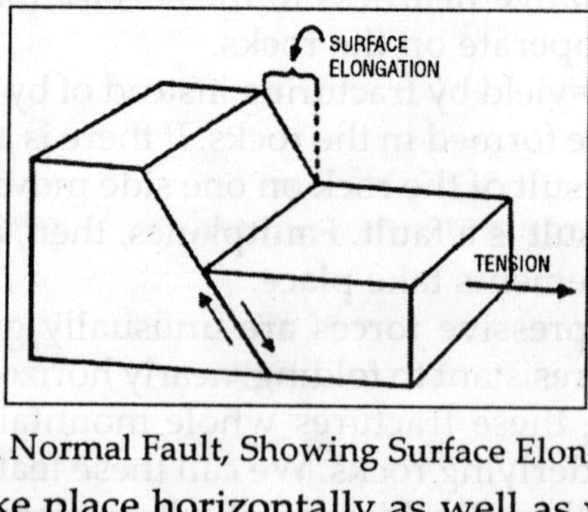

Fig. A Normal Fault, Showing Surface Elongation

Faults may take place horizontally as well as vertically. They, in common with folds, may display any degree of complexity. There are some regions so intricately faulted that they defy full understanding.

There also are regions in which folds and faults are so intimately intermingled that the task of untangling them is wellnigh hopeless. Futhermore, folded and faulted regions may later be subjected to more folding or faulting or both, and the final picture may be one that gives nightmares even to a specialist.

Now while folds and faults are the two main effects of diastrophism in a broad sense, there are other effects in detail. Thus chemical rearrangements that we mentioned in passing have profound effects on the nature of the rocks. New minerals are often formed, and the rocks take on new appearances to such an extent that we consider them a separate group of rocks. They are called the metamorphic rocks, either of the two broad classes of rocks, igneous and sedimentary, may be altered to metamorphic rocks.

He must not gain the impression that mountains are folded up over night. No, it is an extremely slow process, as most geological phenomena are. We have pretty good evidence for believing that some mountain ranges are today in process of growing, but of course that growth is not revealed to a casual glance. While folding may be extremely slow in its development, that need not necessarily be true of faulting.

In fact, we know it isn't true, because faults often take place with enough suddenness to set up distinct shocks in the earth. When these shocks occur, the region adjacent to the fault suffers an earthquake. So important are earthquakes to mankind, both because of their destructiveness and their scientific value, that we shall introduce them to the reader in some detail.

EARTHQUAKES

Among all the cataclysms of nature, none is more terrifying than an earthquake. It has been said that fear of earth tremors increases with increasing experience, whereas in almost any other peril familiarity breeds a certain degree of indifference. Whatever the reason for this, assuming it to be true, there doubtlessly is involved our unconscious acceptance of the earth as a firm and unyielding foundation.

When even this moves beneath our feet, it is not surprising that one's consternation should exceed his judgment. Combine with this the jumble of shattered houses, of gaps and fissures that open in the earth, of springs and streams gone suddenly dry, the low rumble of earth noises, and the dust, confusion, and conflagration that are summed in the word *earthquake,* and the cumulative effect is hardly surprising.

So common are earthquakes that the instruments used to record them show several thousand a year. Most of these are imperceptible tremors, of course; but it must be true that somewhere on the earth every day some people are subjected to greater or less disturbances. How is it, then, that so few of us have ever experienced an earthquake?

Simply because earthquakes, despite their abundance, occur in certain restricted areas on the earth and because, beyond a relatively small radius, the effects of the shocks are not perceptible to our senses. Records covering many years show that most earthquakes are confined to certain belts about the earth. Many of these belts are close to continental margins.

The accompanying map indicates the extent of these belts; and if you will examine it in detail you will realise that practically every earthquake you have ever read about has occurred at some locality within these zones. We cannot say that no earthquake will ever occur outside of these zones; but beyond the limits of the heavy belts, earthquake insurance is pretty much of an idle gesture.

A great deal of evidence collected from many parts of the world shows quite conclusively that most major earthquakes are due to faulting. Quite commonly fissures are formed in the ground during major earthquakes, and these fissures are found to be the surface expressions of faults.

The movement along fault planes may be relatively sudden, and this movement sets up powerful vibrations, which give rise to waves that radiate outward from the centre of the disturbance. It is the sudden shock of these waves that does some of the damage.

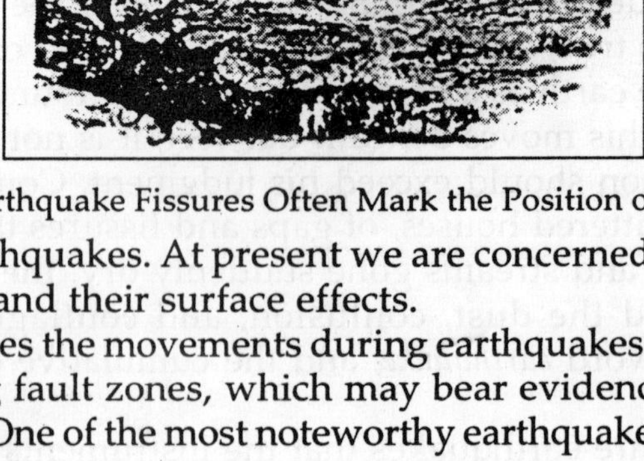

Fig. Earthquake Fissures Often Mark the Position of Faults

During earthquakes. At present we are concerned with the causes of earthquakes and their surface effects.

In many cases the movements during earthquakes take place along already existing fault zones, which may bear evidences of numerous displacements. One of the most noteworthy earthquakes of recent times was that of San Francisco in April, 1906. This earthquake resulted from a displacement of the rocks along the nearly vertical *San Andreas fault*, a zone of weakness that can be traced for many miles through California.

The maximum movement along the fault was 21 ft. in a horizontal direction, as shown by displaced fences, roads, and the like. Faults, we may recall, can take place in any direction, depending on local conditions, although it is true that vertical movements are more common than horizontal ones.

Rocks, contrary to popular fancy, are quite elastic. When an external force is applied, the rocks suffer an elastic deformation, or strain. When the strain due to these forces becomes too great, the rock yields. Near the surface the yield is quite sudden and may result in fractures or actual movement along fault planes. The conditions leading up to an earthquake may then be pictured as the gradual accumulation of stresses within the rocks until they are so great that the rock must yield.

Imagine the sides of a fault tightly pressed together, so that movement can only take place through the overcoming of tremendous frictional resistance. If forces tending to produce movement along the fault plane are set up, the rock resists the movement until its elasticity is exceeded. Then a sudden movement takes place, and the vibrations set up during these movements may be quite intense.

This interpretation of earthquakes has direct evidence to support it. Certain slow displacements of landmarks, demanded by the theory, have been reported along the San Andreas fault; and such slow displacements represent the accumulation of stresses within the rocks adjacent to the fault, so that, when the release occurs, the rocks "snap" into their new positions by an *elastic rebound*.

The accompanying three block diagrams represent the picture as a whole. In the first block a pre-existing fault plane is shown. As forces accumulate in the zone adjacent to the fault, we may visualize a tendency on the part of the rock to adjust itself by a slight deformation, as the second block fancifully suggests. No slip has yet occurred along the fault plane, however, but as the elastic limit is reached, the rocks yield by relatively sudden or jerky movements, and the earthquake damage is done.

The third block pictures this final stage, after the faulting has occurred. Once relief is afforded, the stage is set for a repetition of the process. These examples refer to an earthquake in which the movement is horizontal, but obviously a similar set of conditions could apply to movement in a vertical direction.

The theory just presented is relatively recent; the whole subject of earthquake study is still rather new, but the mass of data now available fairly well establishes the broad outlines of the subject. In addition to direct field observations, scientists nowadays have instruments of such sensitivity and precision that even the feeblest earth tremors are faithfully recorded, and cases are not unusual where the centre of the disturbance is determined from these records before even the news reports tell of the damage.

It was not always thus, of course; until shortly before the turn of the present century, the study of earthquake phenomena was confined largely to qualitative observations. Some of the data obtained from these earlier investigations are interesting and important. The exact geological significance of earthquakes was but dimly realized, also, and even today there is much that remains to be explained.

One thing is certain, however: the advent of modern seismology, or the scientific study of earthquakes, has opened up a field which has shed important light on the question of the nature of the earth's interior. More of this anon; at present suppose we journey backward in time about a century and pick the story up there.

In 1846, Robert Mallet, an Englishman, published a memoir on the dynamics of earthquakes. Here, for the first time, were applied the laws of wave motion in solids to the phenomena of earthquakes. In England, where Mallet lived, however, earthquakes are rarities, and so his experience was more theoretical than actual It is reported that when he was awakened one night by the tremor that stirred Britain in 1852, he did not recognize it as an earthquake. Despite this lack of actual experience, Mallet believed that he held the key to many earthquake secrets. His opportunity came late in 1857, when the great Neapolitan earthquake occurred.

So great was the destruction in this Italian earthquake that news of the catastrophe did not reach the outer world until about a week later. Mallet at once proceeded to the stricken area and began his researches. His principal aims were to determine the exact centre of the earthquake disturbance and to find out how deep within the earth the disturbance occurred, for Mallet argued that both of these ends could be gained by a study of the surface effects of the quake.

You undoubtedly know from newspaper accounts that an earthquake may first present itself as a rather sudden shock, accompanied by billowy movements, as though waves were passing along the surface of the ground. The very suddenness of the shock may contribute in part to the damage, although the amplitude of the earthquake waves, the duration of the quake, and resonance in building structures all fit into the complete picture. The general result is that after an earthquake of destructive violence, there usually is a centre of maximum damage, surrounded by areas in which the destruction diminishes.

The sequence ranges from towns that witness the dramatic events sketched in our opening paragraph to more distant towns and villages in which the greatest damage is perhaps the stopping of clocks or the rattling of a few. dishes. Naturally enough, any disturbance originating at some centre dies down as it moves outward, much as the ripples in a pond diminish in size as they move away from the point where the stone is dropped in.

Now, what Mallet did was to set up a scale of damage suffered, a comparative scale of the intensity of the quake at anypoint. A tour of inspection of the towns in the stricken area then afforded him data for drawing on a map a series of curves which enclosed areas of equal damage.

These lines are called isoseismals, and they enclosed elliptical areas increasing in size as he moved away from the centre of greatest damage. The innermost isoseismal included the zone in which the towns were prostrated, the next larger one those where major parts of the towns were destroyed, and so on with decreasing effect outward.

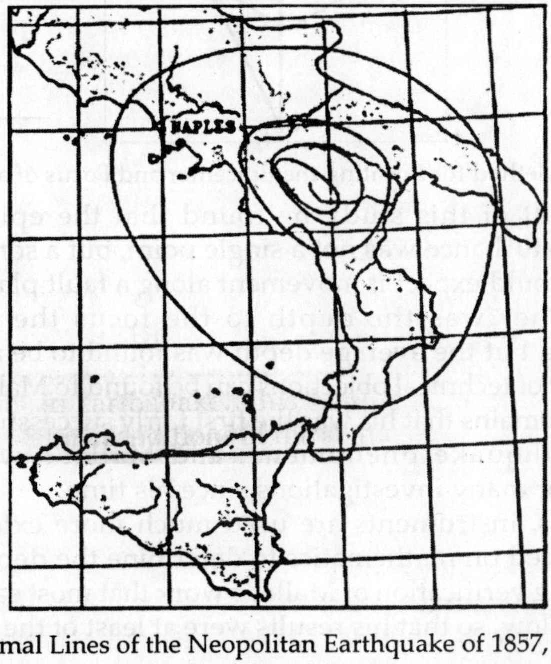

Fig. Isoseismal Lines of the Neopolitan Earthquake of 1857, after Mallet

For locating the centre of the disturbance and its depth, below the surface, Mallet devised an ingenious method. Suppose a disturbance occurs beneath the surface at 0 (called the focus of the earthquake), situated a distance OA beneath the surface. Waves are sent out in all directions from O, and we shall consider one of them, OB.

This strikes the surface at B, at an angle b with the surface. This angle Mallet called the *angle of emergence*. Also, if B is connected with A, the line AB has some given Compass direction; that is, it has an azimuth a with respect to the line AN, supposing N to be the north point.

Now Mallet argued that the direction of travel of the earth wave OB, its angle of emergence b, and the direction AB should be determinable from the effects of the shock at B. The cracks in damaged buildings or walls should be at right angles to the wave direction OB; and overthrown monuments and the like should be found lying in the direction BC, with their bases toward the point A, which is called the epicenter of the earthquake.

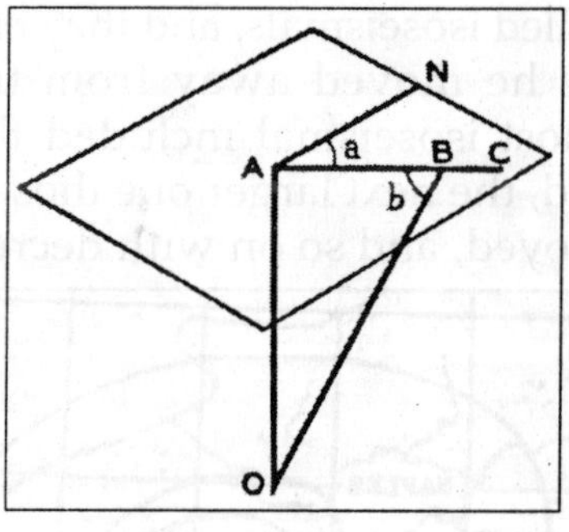

Fig. Mallet's Method for Locating the Epicenter and Focus of an Earthquake

As a result of this study he found that the epicenter of the Neapolitan disturbance was not a single point, but a somewhat linear zone (as we would expect if movement along a fault plane caused the quake). Neither was the depth to the focus the same for all measurements, but the average depth was found to be about 6½ mi.

A number of technical objections can be found to Mallet's methods; but the point remains that his was the first fairly successful quantitative study of earthquake phenomena, and Mallet's work laid the cornerstone for many investigations since his time.

Nowadays, instruments are used much more extensively, and reliance is placed on mathematics to determine the depth of focus. It is an interesting verification of Mallet's work that most earthquake foci are rather shallow, so that his results were at least of the right order of magnitude.

Despite the shift of emphasis toward more instrumental methods, there is no doubt that Mallet gave a decided impetus to earthquake study, and demonstrated that earthquakes, in common with most other earthly phenomena, were amenable to investigation by observation and inference.

As time went on, Mallet's methods were extended and improved upon; and with a growing realization of the nature of the disturbances, instruments were devised for recording many features of quakes that could not be determined from the surface effects of the quake itself in the stricken area. The great advantage of instrumental observation is that it obviates the necessity of being on the ground during or after the earthquake; the instrument may be located far from the disturbance, and yet from the record it traces the distance to the centre of disturbance and the violence of the shock can be determined.

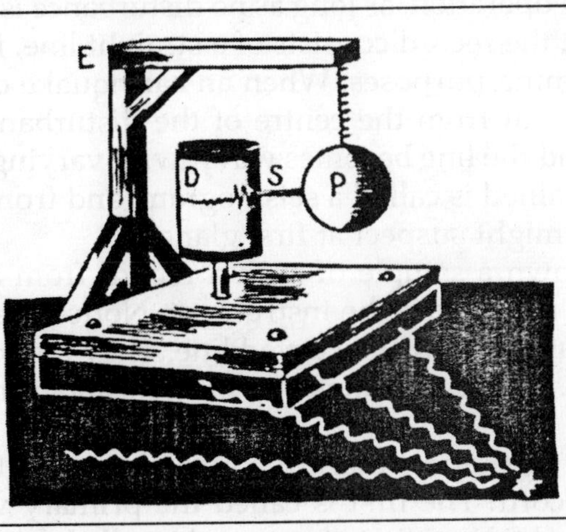

Fig. Simplified Diagram of a Vertical Component Seismograph

The instruments used to detect earthquake waves are called seismographs, and in principle they are quite simple. A seismograph is essentially a pendulum with a long period. A much simplified diagram of one is shown herewith. A heavy weight, P, is suspended by a spring from a rigid frame, E, firmly fastened to the earth, often by means of a pillar of concrete resting directly on bedrock.

On a platform of this frame is a revolving drum, D. A stylus, S, attached to the weight, rests against a sheet of paper wound around the drum. Now imagine an earthquake suddenly vibrating the frame E, The frame, including the drum, will then oscillate a bit, while the mass P, because of its great inertia, will momentarily remain at rest.

Thus a wavy line will be traced on the paper, recording the passage of the earth wave. This is called a *vertical-component seismograph*; and in practice two instruments are used, one to record the vertical components of the wave and the other to detect the horizontal components.

Actually, of course, modern seismographs are much more refined and complicated than this. Timing devices are put on, so that the exact moment of the wave's passing is recorded, and usually the record is made by a beam of light on photographic paper instead of with a stylus. Further, the natural vibrations of the pendulum must be severely damped, so that it does not unduly distort the earthquake waves with its own periodic motion.

We shall not bother the reader with these technical details, but specialists must take careful account of them in actual practice. Plate 24 illustrates modern instruments and their records.

In ordinary operation, as long as no disturbance is passing through the instrument, the record consists of a straight line, interrupted once a minute for timing purposes. When an earthquake occurs, however, the waves sent out from the centre of the disturbance pass through the machine, and the line becomes wavy, with varying amplitude. The record thus obtained is called a seismogram, and from it is read much more than one might suspect at first glance.

Here is shown a sample of such a record from an earthquake a thousand or so miles from the instrument. Notice the letters p, s, and l that are indicated below the curved line. They mark the appearance of the three wave trains that make up the a seismogram, or earthquake record.

The letters p, s, and l mark the arrival-time of the three sets of waves usual record. The first is called the primary wave, because it travels fastest and arrives first; the second is called the secondary wave, and the third is called the long wave. The exact nature of these waves will be developed in our sequel because they afford such an interesting example of agreement between theory and practice.

Aside from any theory, however, it was noted that these waves are separated from each other on the seismogram by time intervals which increase as the distance from the disturbance to the observatory increases. Thus, at an Observatory a few hundred miles from the earthquake the times of arrival of the waves may differ by only a few minutes, whereas from more distant earthquakes they are separated by greater and greater intervals.

An important practical application is made of this timelag between the earthquake waves. It was found by plotting the data of many seismograms that there is a definite relation between the time of arrival of the p waves and s waves, for example, and the distance from the

Observatory to the quake. This relation holds regardless of the geographic location of the earthquake or of the observatory; in short, the time-lag is a function of distance only. Hence, to determine the distance between the observatory and the disturbance, it is only necessary to observe the time in minutes, say, that elapses between the arrival of the first two waves.

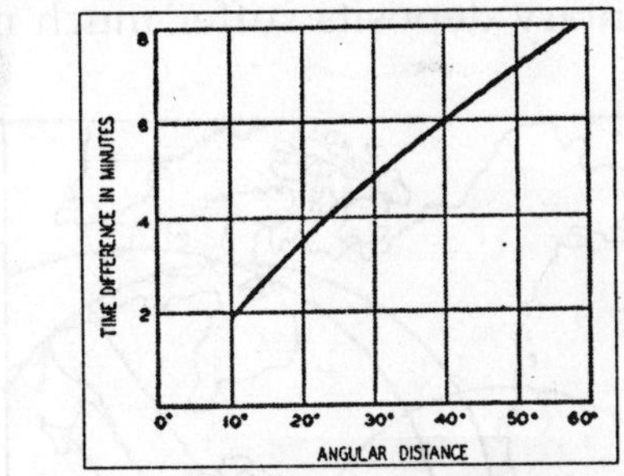

Fig. The Relation Between Distance, and the Difference in Time of Arrival of the p and s Waves

Graphs have been prepared to show this relation between timelag and distance, and the adjacent diagram is one somewhat simplified. The time difference in minutes between the arrival of the p and s waves is shown along the vertical axis, and the angular distance to the epicenter is given along the horizontal axis.

Suppose, for example, that the p wave is recorded on the seismogram at 7:02 A.M. and the s wave at 7:07 A.M. Here is a difference of 5 min.; and by finding this value on the vertical axis, the corresponding distance is seen to be 30°, or about 2,100 mi.

Now, for any one station to determine the *direction* to the epicenter is somewhat tedious. It requires quite a bit of calculation, so that several observatories usually co-operate in locating the stricken area, for, notice that if each station can determine the distance, then three stations working together may locate the centre of the shock. This is how it is done.

A circle is drawn about each observatory as a centre, with a radius equal to the computed distance to the quake. The intersection of the three circles is the desired location. The adjoining figure illustrates how observatories at Pasadena, Chicago, and Washington may locate an epicenter in Mexico.

The example just given serves as a striking contrast to the much less refined methods that were available in Mallet's time. During the earthquake of 1857, the area was so stricken that news of the quake was delayed nearly a week; nowadays the recording and location of quakes are a matter of hours only. Earthquake waves travel at the rate of several miles per second, so that almost immediately after the shock seismographs for many miles around record the disturbance.

Not only have methods of studying earthquakes improved tremendously in the last third- or half-century, but much more is now known about the immediate causes of the destructive effects. It has been amply demonstrated, for example, that structures on mantle rock and on loose sedimentary deposits suffer much more than those on solid bedrock.

Fig. Three Observatories Co-Operate in Locating a Quake in Mexico

In the great San Francisco quake of 1906 it was the buildings erected on filled-in land that suffered most; the same was true of the Messina earthquake of 1908 and the Tokio quake of 1923.

From an engineering viewpoint, then, earthquake research has done much to avoid the extreme destruction of earlier shocks, both by new types of shock-resisting construction and by recognizing the probable centers of greatest damage in regions subject to earthquakes.

The reason why earthquake damage is greatest on loose or unconsolidated rock is not hard to understand.

As long as the earthquake waves pass through dense rock, they merely vibrate the particles; but when they enter loose materials, their energy may be transformed to an actual shifting of the particles among themselves.

We may illustrate this by placing a marble on the floor. If the floor is tapped smartly with a hammer several feet away, no movement of the floor is perceptible, but the marble is made to jump several inches into the air. The energy here has been partly applied to movement of the loose marble itself.

Inasmuch as earthquake waves are partly transmitted beneath the earth's surface during their propagation from the focus, we may expect that a theoretical study of the waves themselves could shed light on the nature of the earth's interior. This is exactly the case: earthquake waves are our most powerful weapon for studying the deep interior of the earth.

Chapter 12

Causes of Earthquake

The sudden movement of the Earth caused by the abrupt release of accumulated strain along a fault in the interior. The released energy passes through the Earth as seismic waves (low-frequency sound waves), which cause the shaking. Seismic waves continue to travel through the Earth after the fault motion has stopped. Recordings of earthquakes, called seismograms, illustrate that such motion is recorded all over the Earth for hours, and even days, after an earthquake.

Earthquakes are not distributed randomly over the globe but tend to occur in narrow, continuous belts of activity. Approximately 90% of all earthquakes occur in these belts, which define the boundaries of the Earth's plates. The plates are in continuous motion with respect to one another at rates on the order of centimeters per year; this plate motion is responsible for most geological activity.

Plate motion occurs because the outer cold, hard skin of the Earth, the lithosphere, overlies a hotter, soft layer known as the asthenosphere. Heat from decay of radioactive minerals in the Earth's interior sets the asthenosphere into thermal convection. This convection has broken the lithosphere into plates which move about in response to the convective motion.

As the plates move past each other, little of the motion at their boundaries occurs by continuous slippage; most of the motion occurs in a series of rapid jerks. Each jerk is an earthquake. This happens because, under the pressure and temperature conditions of the shallow part of the Earth's lithosphere, the frictional sliding of rock exhibits a property known as stick-slip, in which frictional sliding occurs in a series of jerky movements, interspersed with periods of no motion—or sticking. In the geologic time frame, then, the lithospheric plates chatter at their boundaries, and at any one place the time between chatters may be hundreds of years.

The periods between major earthquakes is thus one during which strain slowly builds up near the plate boundary in response to the continuous movement of the plates. The strain is ultimately released by an earthquake when the frictional strength of the plate boundary is exceeded.

Most great earthquakes occur on the boundaries between lithospheric plates and arise directly from the motions between the plates. These may be called plate boundary earthquakes. There are many earthquakes, sometimes of substantial size, that cannot be related so simply to the movements of the plates. At many plate boundaries, earthquakes occur over a broad zone—often several hundred miles wide—adjacent to the plate boundary.

These earthquakes, which may be called plate boundary-related earthquakes, are secondarily caused by the stresses set up at the plate boundary. Some earthquakes also occur, although infrequently, within plates. These earthquakes, which are not related to plate boundaries, are called intraplate earthquakes. The immediate cause of intraplate earthquakes is not understood.

In addition to the tectonic types of earthquakes described above, some earthquakes are directly associated with volcanic activity. These volcanic earthquakes result from the motion of undergound magma that leads to volcanic eruptions.

Earthquakes often occur in well-defined sequences in time. Tectonic earthquakes are often preceded, by a few days to weeks, by several smaller shocks (foreshocks), and are nearly always followed by large numbers of aftershocks. Foreshocks and aftershocks are usually much smaller than the main shock. Volcanic earthquakes often occur in flurries of activity, with no discernible main shock. This type of sequence is called a swarm.

Earthquakes range enormously in size, from tremors in which slippage of a few tenths of an inch occurs on a few feet of fault, to the greatest events, which may involve a rupture many hundreds of miles long, with tens of feet of slip.

The size of an earthquake is given by its moment: average slip times the fault area that slipped times the elastic constant of the Earth. The units of seismic moment are dyne-centimeters. An older measure of earthquake size is magnitude, which is proportional to the logarithm of moment. Magnitude 2.0 is about the smallest tremor that can be felt.

Most destructive earthquakes are greater than magnitude 6; the largest shock known was the 1960 Chile earthquake, with a moment of 10^{30} dyne-centimeters (10^{23} newton-meters) or magnitude 9.5. It involved a fault 600 mi (1000 km) long slipping 30 ft (10 m).

The intensity of an earthquake is a measure of the severity of shaking and its attendant damage at a point on the surface of the Earth.

The same earthquake may therefore have different intensities at different places. The intensity usually decreases away from the epicenter (the point on the surface directly above the onset of the earthquake), but its value depends on many factors and generally increases with moment. Intensity is usually higher in areas with thick alluvial cover or landfill than in areas of shallow soil or bare rock.

Poor building construction leads to high intensity ratings because the damage to structures is high. Intensity is therefore more a measure of the earthquake's effect on humans than an innate property of the earthquake.

An earthquake is the result of a sudden release of energy in the Earth's crust that creates seismic waves. Earthquakes are recorded with a seismometer, also known as a seismograph. The moment magnitude of an earthquake is conventionally reported, or the related and mostly obsolete Richter magnitude, with magnitude 3 or lower earthquakes being mostly imperceptible and magnitude 7 causing serious damage over large areas. Intensity of shaking is measured on the modified Mercalli scale.

At the Earth's surface, earthquakes manifest themselves by shaking and sometimes displacing the ground. When a large earthquake epicenter is located offshore, the seabed sometimes suffers sufficient displacement to cause a tsunami. The shaking in earthquakes can also trigger landslides and occasionally volcanic activity.

In its most generic sense, the word *earthquake* is used to describe any seismic event—whether a natural phenomenon or an event caused by humans—that generates seismic waves. Earthquakes are caused mostly by rupture of geological faults, but also by volcanic activity, landslides, mine blasts, and nuclear experiments. An earthquake's point of initial rupture is called its focus or hypocenter. The term epicenter refers to the point at ground level directly above this.

NATURALLY OCCURRING EARTHQUAKES

Tectonic earthquakes will occur anywhere within the earth where there is sufficient stored elastic strain energy to drive fracture propagation along a fault plane. In the case of transform or convergent type plate boundaries, which form the largest fault surfaces on earth, they will move past each other smoothly and aseismically only if there are no irregularities or asperities along the boundary that increase the frictional resistance. Most boundaries do have such asperities and this leads to a form of stick-slip behaviour.

Once the boundary has locked, continued relative motion between the plates leads to increasing stress and therefore, stored strain energy in the volume around the fault surface. This continues until the stress has risen sufficiently to break through the asperity, suddenly allowing sliding over the locked portion of the fault, releasing the stored energy. This energy is released as a combination of radiated elastic strain seismic waves, frictional heating of the fault surface, and cracking of the rock, thus causing an earthquake.

This process of gradual build-up of strain and stress punctuated by occasional sudden earthquake failure is referred to as the Elastic-rebound theory. It is estimated that only 10 percent or less of an earthquake's total energy is radiated as seismic energy. Most of the earthquake's energy is used to power the earthquake fracture growth or is converted into heat generated by friction.

Therefore, earthquakes lower the Earth's available elastic potential energy and raise its temperature, though these changes are negligible compared to the conductive and convective flow of heat out from the Earth's deep interior.

CAUSES OF EARTHQUAKE

An Earthquake is a sudden tremor or movement of the earth's crust, which originates naturally at or below the surface. The word natural is important here, since it excludes shock waves caused by French nuclear tests, man made explosions and landslides caused by building work.

There are two main causes of earthquakes:

- Firstly, they can be linked to explosive volcanic eruptions; they are in fact very common in areas of volcanic activity where they either proceed or accompany eruptions.
- Secondly, they can be triggered by Tectonic activity associated with plate margins and faults. The majority of earthquakes world wide are of this type.

An earthquake can be likened to the effect observed when a stone is thrown into water. After the stone hits the water a series of concentric waves will move outwards from the centre. The same events occur in an earthquake. There is a sudden movement within the crust or mantle, and concentric shock waves move out from that point.

Geologists and Geographers call the origin of the earthquake the focus. Since this is often deep below the surface and difficult to map, the location of the earthquake is often referred to as the point on the Earth surface directly above the focus. This point is called the epicentre.

The strength, or magnitude, of the shockwaves determines the extent of the damage caused. Two main scales exist for defining the strength, the Mercalli Scale and the Richter Scale.

Earthquakes are three dimensional events, the waves move outwards from the focus, but can travel in both the horizontal and vertical plains. This produces three different types of waves which have their own distinct characteristics and can only move through certain layers within the Earth. Lets take a look at these three forms of shock waves.

TYPES OF SHOCKWAVES

P-Waves

Primary Waves (P-Waves) are identical in character to sound waves. They are high frequency, short-wavelength, longitudinal waves which can pass through both solids and liquids. The ground is forced to move forwards and backwards as it is compressed and decompressed. This produces relatively small displacements of the ground.

P Waves can be reflected and refracted, and under certain circumstances can change into S-Waves.

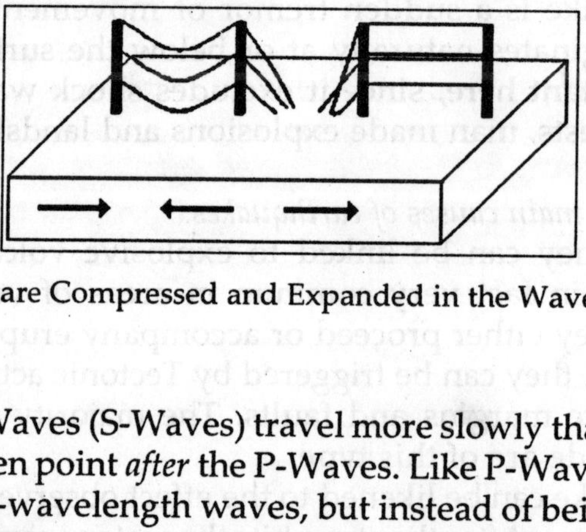

Fig. Particles are Compressed and Expanded in the Wave's Direction.

S-Waves

Secondary Waves (S-Waves) travel more slowly than P-Waves and arrive at any given point *after* the P-Waves. Like P-Waves they are high frequency, short-wavelength waves, but instead of being longitudinal they are transverse.

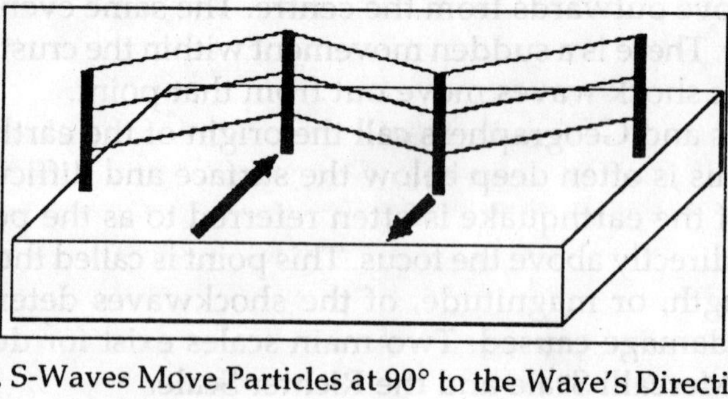

Fig. S-Waves Move Particles at 90° to the Wave's Direction.

They move in all directions away from their source, at speeds which depend upon the density of the rocks through which they are moving. They cannot move through liquids. On the surface of the Earth, S-Waves are responsible for the sideways displacement of walls and fences, leaving them 'S' shaped.

L-Waves

Surface Waves (L-Waves) are low frequency transverse vibrations with a long wavelength. They are created close to the epicentre and can only travel through the outer part of the crust. They are responsible for the majority of the building damage caused by earthquakes.

This is because L Waves have a motion similar to that of waves in the sea. The ground is made to move in a circular motion, causing it to rise and fall as visible waves move across the ground. Together with secondary effects such as landslides, fires and tsunami these waves account for the loss of approximately 10,000 lives and over $100 million per year.

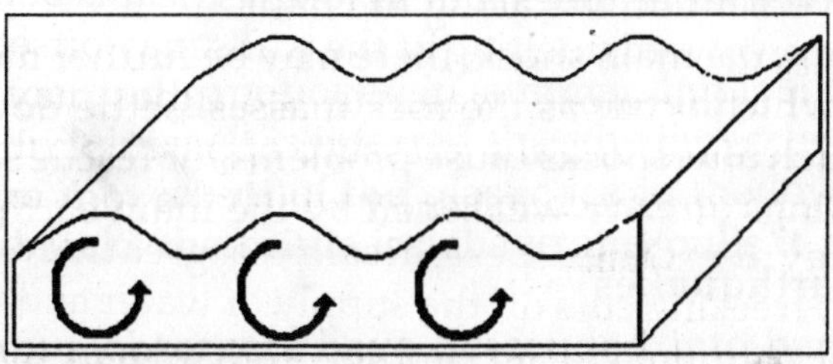

Fig. L-Waves Move Particles in a Circular Path.

Tectonic Earthquakes

Tectonic earthquakes are triggered when the crust becomes subjected to strain, and eventually moves. The theory of plate tectonics explains how the crust of the Earth is made of several plates, large areas of crust which float on the Mantle.

Since these plates are free to slowly move, they can either drift towards each other, away from each other or slide past each other. Many of the earthquakes which we feel are located in the areas where plates collide or try to slide past each other.

The process which explains these earthquakes, known as *Elastic Rebound Theory* can be demonstrated with a green twig or branch. Holding both ends, the twig can be slowly bent. As it is bent, energy is built up within it. A point will be reached where the twig suddenly snaps. At this moment the energy within the twig has exceeded the *Elastic Limit* of the twig. As it snaps the energy is released, causing the twig to vibrate and to produce sound waves. Perhaps the most famous example of plates sliding past each other is the San Andreas Fault in California.

Here, two plates, the Pacific Plate and the North American Plate, are both moving in a roughly northwesterly direction, but one is moving faster than the other. The San Francisco area is subjected to hundreds of small earthquakes every year as the two plates grind against each other. Occasionally, as in 1989, a much larger movement occurs, triggering a far more violent 'quake'.

Major earthquakes are sometimes preceded by a period of changed activity. This might take the form of more frequent minor shocks as the rocks begin to move,called *foreshocks,* or a period of less frequent shocks as the two rock masses temporarily 'stick' and become locked together.

Detailed surveys in San Francisco have shown that railway lines, fences and other longitudinal features very slowly become deformed as the pressure builds up in the rocks, then become noticeably offset when a movement occurs along the fault.

Following the main shock, there may be further movements, called *aftershocks,* which occur as the rock masses 'settle down' in their new positions. Such aftershocks cause problems for rescue services, bringing down buildings already weakened by the main earthquake.

Volcanic Earthquakes

Volcanic earthquakes are far less common than Tectonic ones. They are triggered by the explosive eruption of a volcano. Given that not all volcanoes are prone to violent eruption, and that most are 'quiet' for the majority of the time, it is not surprising to find that they are comparatively rare.

When a volcano explodes, it is likely that the associated earthquake effects will be confined to an area 10 to 20 miles around its base, where as a tectonic earthquake may be felt around the globe.

The volcanoes which are most likely to explode violently are those which produce *acidic* lava. Acidic lava cools and sets very quickly upon contact with the air. This tends to chock the volcanic vent and block the further escape of pressure. For example, in the case of Mt Pelee, the lava solidified before it could flow down the sides of the volcano. Instead it formed a spine of solid rock within the volcano vent.

The only way in which such a blockage can be removed is by the build up of pressure to the point at which the blockage is literally exploded out of the way. In reality, the weakest part of the volcano will be the part which gives way, sometimes leading to a sideways explosion as in the Mt St.Helens eruption. When extraordinary levels of pressure develop, the resultant explosion can be devastating, producing an earthquake of considerable magnitude. When Krakatoa (Indonesia, between Java and Sumatra) exploded in 1883, the explosion was heard over 5000 km away in Australia. The shockwaves produced a series of tsunami (large sea waves), one of which was over 36m high; that's the same as four, two story houses stacked on top of each other. These swept over the coastal areas of Java and Sumatra killing over 36,000 people.

By contrast, volcanoes producing free flowing *basic* lava rarely cause earthquakes. The lava flows freely out of the vent and down the sides of the volcano, releasing pressure evenly and constantly. Since pressure doesn't build up, violent explosions do not occur.

FAULT

In geology, a fault or fault line is a planar fracture in rock in which the rock on one side of the fracture has moved with respect to the rock on the other side. Large faults within the Earth's crust are the result of differential or shear motion and active fault zones are the causal locations of most earthquakes. Earthquakes are caused by energy release during rapid slippage along a fault. A fault that runs along the boundary between two tectonic plates is called a transform fault.

Since faults do not usually consist of a single, clean fracture, the term *fault zone* is used when referring to the zone of complex deformation that is associated with the fault plane.

The two sides of a non-vertical fault are called the *hanging wall* and *footwall*. By definition, the hanging wall occurs above the fault and the footwall occurs below the fault. This terminology comes from mining. When working a tabular ore body the miner stood with the footwall under his feet and with the hanging wall hanging above him.

The creation and behaviour of faults, in both an individual small fault and within the greater fault zones which define the tectonic plates, is controlled by the relative motion of rocks on either side of the fault surface.

Because of friction and the rigidity of the rock, the rocks cannot simply glide or flow past each other. Rather, stress builds up in rocks and when it reaches a level that exceeds the strain threshold, the accumulated potential energy is released as strain, which is focused into a plane along which relative motion is accommodated — the fault.

Strain is both accumulative and instantaneous depending on the rheology of the rock; the ductile lower crust and mantle accumulates deformation gradually via shearing whereas the brittle upper crust reacts by fracture, or instantaneous stress release to cause motion along the fault. A fault in ductile rocks can also release instantaneously when the strain rate is too great. The energy released by instantaneous strain release is the cause of earthquakes, a common phenomenon along transform boundaries.

Mechanics

The creation and behaviour of faults, in both an individual small fault and within the greater fault zones which define the tectonic plates, is controlled by the relative motion of rocks on either side of the fault surface.

Because of friction and the rigidity of the rock, the rocks canno simply glide or flow past each other. Rather, stress builds up in rocks and when it reaches a level that exceeds the strain threshold, the accumulated potential energy is released as strain, which is focused into a plane along which relative motion is accommodated — the fault.

Strain is both accumulative and instantaneous depending on the rheology of the rock; the ductile lower crust and mantle accumulates deformation gradually via shearing whereas the brittle upper crust reacts by fracture, or instantaneous stress release to cause motion along the fault.

A fault in ductile rocks can also release instantaneously when the strain rate is too great. The energy released by instantaneous strain release is the cause of earthquakes, a common phenomenon along transform boundaries.

Microfracturing and AMR Theory

Microfracturing, or microseismicity, is sometimes thought of as a symptom caused by rocks under strain, where small-scale failures, perhaps on areas the size of a dinner plate or a small area, release stress under high strain conditions. It is only when sufficient microfractures link up into a large slip surface that a large seismic event or earthquake can occur.

According to this theory, after a large earthquake, the majority of the stress is released and the frequency of microfracturing is exponentially lower.

A related theory, accelerating moment release (AMR), hypothesizes that the seismicity rate accelerates in a well-behaved way prior to large earthquakes, and may be a promising tool for earthquake prediction on the scale of days to years.

This is being increasingly used to predict rock failures within mines and applications are being attempted for the portions of faults within brittle rheological conditions. Similar behaviour is observed in the tremors preceding volcanic eruptions.

Slip, Heave, Throw

The sense of slip is defined by the relative movements of geological features present on either side of the fault plane and is a vector. The sense of slip defines the type of fault. This is distinct from the throw of the fault, which is the vertical offset. Heave is the measured horizontal offset of the fault.

The vector of slip can be qualitatively measured by fault bend folding, i.e. drag folding of strata on either side of the fault; the direction and magnitude of heave and throw can be measured only by finding common intersection points on either side of the fault. In practise it is usually only possible to find the slip direction of faults, and an approximation of the heave and throw vector.

Fault Types

Faults can be categorized into three groups based on the sense of slip. A fault where the main sense of movement (or slip) on the fault plane is vertical is known as a dip-slip fault. Where the main sense of slip is horizontal the fault is known as a transcurrent or strike-slip fault. Oblique-slip faults have significant components of both strike and dip slip.

For all naming distinctions, it is the orientation of the net dip and sense of slip of the fault which must be considered, not the present-day orientation, which may have been altered by local or regional folding or tilting.

Dip-Slip Faults

Dip-slip faults can be again classified into the types "reverse" and "normal". A normal fault occurs when the crust is extended. Alternatively such a fault can be called an extensional fault. The hanging wall moves downward, relative to the footwall. A downthrown block between two normal faults dipping towards each other is called a graben. An upthrown block between two normal faults dipping away from each other is called a horst. Low-angle normal faults with regional tectonic significance may be designated detachment faults.

A reverse fault is the opposite of a normal fault — the hanging wall moves up relative to the footwall. Reverse faults are indicative of shortening of the crust. The dip of a reverse fault is relatively steep, greater than 45°.

A thrust fault has the same sense of motion as a reverse fault, but with the dip of the fault plane at less than 45°. Thrust faults typically form ramps, flats and fault-bend (hanging wall and foot wall) folds. Thrust faults are responsible for forming nappes and klippen in the large thrust belts.

The fault plane is the plane that represents the fracture surface of a fault. Flat segments of thrust fault planes are known as *flats,* and inclined sections of the thrust are known as *ramps*. Typically thrust faults move *within* formations by forming flats, and climb up section with ramps.

Fault-bend folds are formed by movement of the hangingwall over a non-planar fault surface and are found associated with both extensional and thrust faults.

Faults may be reactivated at a later time with the movement in the opposite direction to the original movement (fault inversion). A normal fault may therefore become a reverse fault and vice versa.

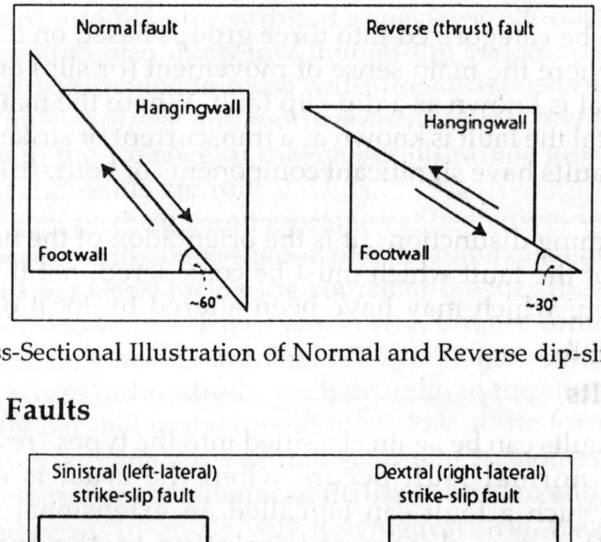

Fig. Cross-Sectional Illustration of Normal and Reverse dip-slip Faults.

Strike-Slip Faults

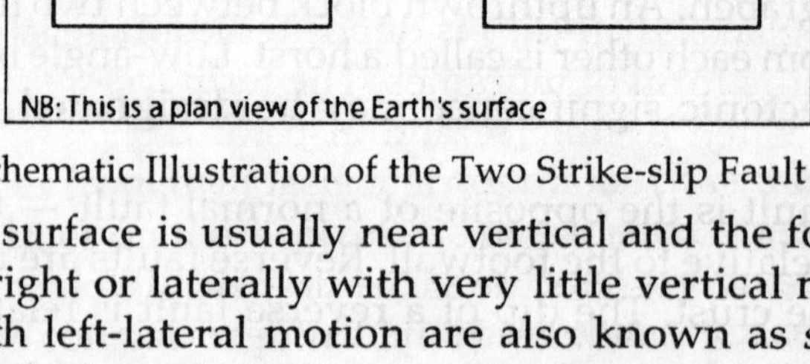

Fig. Schematic Illustration of the Two Strike-slip Fault Types.

The fault surface is usually near vertical and the footwall moves either left or right or laterally with very little vertical motion. Strike-slip faults with left-lateral motion are also known as *sinistral* faults. Those with right-lateral motion are also known as *dextral* faults. A special class of strike-slip faults is the transform faults which are a plate tectonics feature related to spreading centers such as mid-ocean ridges.

Oblique-Slip Faults

A fault which has a component of dip-slip and a component of strike-slip is termed an oblique-slip fault. Nearly all faults will have some component of both dip-slip and strike-slip, so defining a fault as oblique requires both dip and strike components to be measurable and significant. Some oblique faults occur within transtensional and transpressional regimes, others occur where the direction of extension or shortening changes during the deformation but the earlier formed faults remain active.

HADE: Hade is defined as the complement of dip, it is the angle between the fault plane and a vertical plane that strikes parallel to the fault.

Fault Rock

All faults have a measurable thickness, made up of deformed rock that is characteristic of the level in the crust where the faulting happened, the rock types affected by the fault and the presence and nature of any mineralising fluids. Fault rocks are classified by their textures and the implied mechanism of deformation.

A fault that passes through different levels of the lithosphere will have many different types of fault rock developed along its surface. Continued dip-slip displacement will tend to juxtapose fault rocks characteristic of different crustal levels, with varying degrees of overprinting.

This effect is particularly clear in the case of detachment faults and major thrust faults.

The main types of fault rock are:

- *Cataclasite:* A fault rock which is cohesive with a poorly developed or absent planar fabric, or which is incohesive, characterised by generally angular clasts and rock fragments in a finer-grained matrix of similar composition.
- *Mylonite:* A fault rock which is cohesive and characterized by a well developed planar fabric resulting from tectonic reduction of grain size, and commonly containing rounded porphyroclasts and rock fragments of similar composition to minerals in the matrix
- *Tectonic or Fault Breccia:* A medium- to coarse-grained cataclasite containing >30% visible fragments.
- *Fault Gouge:* An incohesive, clay-rich fine- to ultrafine-grained

cataclasite, which may possess a planar fabric and containing <30% visible fragments. Rock clasts may be present

- *Pseudotachylite:* Ultrafine-grained vitreous-looking material, usually black and flinty in appearance, occurring as thin planar veins, injection veins or as a matrix to pseudoconglomerates or breccias, which infills dilation fractures in the host rock.
- *Clay smear clay:* Rich fault gouge formed in sedimentary sequences containing clay-rich layers which are strongly deformed and sheared into the fault gouge.

AFTERSHOCK

An aftershock is an earthquake that occurs after a previous earthquake (the main shock). An aftershock is in the same region of the main shock but is always of smaller magnitude strength. If an aftershock is larger than the main shock, the aftershock is redesignated as the main shock and the original main shock is redesignated as a foreshock. Aftershocks are smaller earthquakes formed as the crust around the displaced fault plane adjusts to the effects of the main shock.

Distribution of Aftershocks

Most aftershocks are located over the full area of fault rupture and either occur along the fault plane itself or along other faults within the volume affected by the strain associated with the main shock. Typically, aftershocks are found up to a distance equal to the rupture length away from the fault plane.

The pattern of aftershocks helps confirm the size of area that slipped during the main shock. In the case of the 2004 Indian Ocean earthquake and the 2008 Sichuan earthquake the aftershock distribution shows that in both cases that the epicenter (where the rupture initiated) lies to one end of the final area of slip, implying strongly asymmetric rupture propagation.

Aftershock Size and Frequency with Time

Aftershocks tend to obey a number of empirical laws concerning magnitude and frequency.

Omori's Law

Aftershocks occur with a pattern that follows Omori's law. Omori's law, or more correctly the modified Omori's law, is an empirical relation for the temporal decay of aftershock rates. In 1894, Omori published his work on the aftershocks of earthquakes, in which he stated that aftershock frequency decreases by roughly the reciprocal of time after the main shock.

$$n(t)=\frac{K}{c+t}$$

where:

- *n(t)* is the rate of earthquakes measured in a certain time *t* after the main shock,
- *K* is the amplitude, and
- *c* is the "time offset" parameter.

The modified version of Omori's law, now commonly used, was proposed by Utsu in 1961.

$$n(t)=\frac{K}{(c+t)^{P}}$$

where

- *p* modifies the decay rate and typically falls in the range 0.7–1.5.

According to these equations, the rate of aftershocks decreases quickly with time. The rate of aftershocks is proportional to the inverse of time since the mainshock.

Thus whatever the odds of an aftershock are on the first day, the second day will have 1/2 the odds of the first day and the tenth day will have approximately 1/10th the odds of the first day (when *p* is equal to 1).

These patterns describe only the mass behaviour of aftershocks; the actual times, numbers and locations of the aftershocks are 'random', while tending to follow these patterns. As this is an empirical law values of the parameters are obtained by fitting to data after the mainshock occurred and they have no physical basis/meaning.

Bath's Law

The other main law describing aftershocks is known as Bath's Law and this states that the difference in magnitude between a main shock and its largest aftershock is approximately constant, independent of the main shock magnitude, typically 1.1-1.2 on the Moment magnitude scale.

Gutenberg-Richter Law

Aftershock sequences also typically follow the Gutenberg-Richter law of scaling, which refers to the relationship between the magnitude and total number of earthquakes in a region in a given time period. In summary, there are more small aftershocks and fewer large aftershocks.

Impact of Aftershocks

Aftershocks are dangerous because they are usually unpredictable, can be of a large magnitude, and can collapse buildings that are damaged from the mainshock. Bigger earthquakes have more and larger aftershocks and the sequences can last for years or even longer especially when a large event occurs in a seismically quiet area; the New Madrid Seismic Zone, where events still follow Omori's law from the mainshocks of 1811–1812. An aftershock sequence is deemed to have ended when the rate of seismicity drops back to a background level; i.e., no further decay in the number of events with time can be detected.

Foreshocks

Foreshocks have been defined as a tremor of at least 2.5 on the Richter-scale that take place prior to a mainshock tremor of 5.4 or greater on the Richter-scale. Many scientists hope to use foreshocks to predict upcoming earthquakes. In particular, the East Pacific Rise transform faults show foreshock activity before the main seismic event. Reviews of data of past events and their foreshocks showed that they have a low number of aftershocks and high foreshock rates compared to continental strike-slip faults.

EARTHQUAKE SWARM

Earthquake swarms are events where a local area experiences sequences of many earthquakes striking in a relatively short period of time. The length of time used to define the swarm itself varies, but the United States Geological Survey points out that an event may be on the order of days, weeks, or months. They are differentiated from earthquakes succeeded by a series of aftershocks by the observation that no single earthquake in the sequence is obviously the main shock. Earthquake swarms are one of the events typically preceding eruptions of volcanoes.

One example is along the Cerro Prieto fault near Mexicali, BC in Mexico where over 500 quakes and aftershocks hit in February, 2008. Another is a swarm that's been dubbed "The Mogul earthquake sequence" that began in February 2008 near Reno, Nevada and continued for several months, ending in November 2008. Between February and April the swarm produced more than 1,000 quakes of small magnitude, although the largest measured 4.7.

Earthquake Storm

An earthquake storm is a recently proposed theory about

earthquakes, where one triggers a series of other large earthquakes—within the same tectonic plate—as the stress transfers along the fault. This is similar to the idea of aftershocks, with the exception that they take place years apart.

These series of earthquakes can devastate entire countries or geographical regions. Possible events may have occurred during the end of the Bronze Age, and the latter part of the Roman Empire. It has been suggested that this is what may be occurring in modern day Turkey.

PLATE TECTONICS

The earth beneath our feet is not dead; it is constantly moving, driven by forces deep in its core. Nor is the planet's crust all of one piece; it is composed of numerous plates, which are moving steadily in relation to one another. This movement is responsible for all manner of phenomena, including earthquakes, volcanoes, and the formation of mountains. All these ideas, and many more, are encompassed in the concept of plate tectonics, which is the name for a branch of geologic and geophysical study and for a powerful theory that unites a vast array of ideas.

Plate tectonics works hand in hand with several other striking concepts and discoveries, including continental drift and the many changes in Earth's magnetic field that have taken place over its history. No wonder, then, that this idea, developed in the 1960s but based on years of research that preceded that era, is described as "the unifying theory of geology."

TECTONICS AND TECTONISM

The lithosphere is the upper layer of Earth's interior, including the crust and the brittle portion at the top of the mantle. Tectonism is the deformation of the lithosphere, and the term tectonics refers to the study of this deformation, including its causes and effects, most notably mountain building. This deformation is the result of the release and redistribution of energy from Earth's core.

The interior of Earth itself is divided into three major sections: the crust, mantle, and core. The first is the uppermost division of the solid earth, representing less than 1% of its volume and varying in depth from 3 mi. to 37 mi. (5-60 km). Below the crust is the mantle, a thick, dense layer of rock approximately 1,429 mi. (2,300 km) thick.

The core itself is even more dense, as illustrated by the fact that it constitutes about 16% of the planet's volume and 32% of its mass. Composed primarily of iron and another, lighter element (possibly sulfur), it is divided between a solid inner core with a radius of about 760 mi. (1,220 km) and a liquid outer core about 1,750 mi. (2,820 km) thick.

Tectonism results from the release and redistribution of energy from Earth's interior. There are two components of this energy: gravity, a function of the enormous mass at the core, and heat from radioactive decay. Differences in mass and heat within the planet's interior, known as pressure gradients, result in the deformation of rocks.

DEFORMATION OF ROCKS

Any attempt to deform an object is referred to as stress, and stress takes many forms, including tension, compression, and shear. Tension acts to stretch a material, whereas compression—a type of stress produced by the action of equal and opposite forces, whose effect is to reduce the length of a material—has the opposite result. (Compression is a form of pressure.)

As for shear, this is a kind of stress resulting from equal and opposite forces that do not act along the same line. If a thick, hardbound book is lying flat and one pushes the front cover from the side so that the covers and pages are no longer perfectly aligned, this is an example of shear.

Under the effects of these stresses, rocks may bend, warp, slide, or break. They may even flow, as though they were liquids, or melt and thus truly become liquid. As a result, Earth's interior may manifest faults, or fractures in rocks, as well as folds, or bends in the rock structure. The effects of this activity can be seen on the surface in the form of subsidence, which is a depression in the crust, or uplift, which is the raising of crustal materials. Earthquakes and volcanic eruptions also may result.

There are two basic types of tectonism: orogenesis and epeirogenesis. *Orogenesis* is taken from the Greek words *oros* ("mountain") and *genesis* ("origin") and involves the formation of mountain ranges by means of folding, faulting, and volcanic activity. The Greek word *epeiros* means "mainland," and epeirogenesis takes the form of either uplift or subsidence. Of principal concern in the theory of plate tectonics, as we shall see, is orogenesis, which involves more lateral, as opposed to vertical, movement.

CONTINENTAL DRIFT

If one studies a world map for a period of time, one may notice something interesting about the shape of Africa's west coast and that of South America's east coast: they seem to fit together like pieces of a jigsaw puzzle. Early in the twentieth century, two American geologists, Frank Bursley Taylor and Howard Baker, were among the first scientists to point out this fact.

According to Taylor and Baker, Europe, the Americas, and Africa all had been joined at one time. This was an early version of continental drift, a theory concerning the movement of Earth's continents.

Continental drift is based on the idea that the configuration of continents was once different than it is today, that some of the individual landmasses of today once were joined in other continental forms, and that the landmasses later moved to their present locations.

Though Taylor and Baker were early proponents, the theory is associated most closely with the German geophysicist and meteorologist Alfred Wegener, who made the case for continental drift in *The Origin of Continents and Oceans*.

PANGAEA, LAURASIA, AND GONDWANALAND

According to Wegener, the continents of today once formed a single supercontinent called Pangaea, from the Greek words *pan* ("all") and *gaea* ("Earth"). Eventually, Pangaea split into two halves, with the northern continent of Laurasia and the southern continent of Gondwanaland, sometimes called Gondwana, separated by the Tethys Sea. In time, Laurasia split to form North America, the Eurasian landmass with the exception of the Indian subcontinent, and Greenland. Gondwanaland also split, forming the major southern landmasses of the world: Africa, South America, Antarctica, Australia, and India.

The Austrian geologist Eduard Suess and the South African geologist Alexander du Toit, each of whom contributed significantly to continental drift theory, were responsible for the naming of Gondwanaland and Laurasia, respectively. Suess preceded Wegener by many years with his theory of Gondwanaland, named after the Gondwana region of southern India.

There he found examples of a fern that, in fossilized form, had been found in all the modern-day constituents of the proposed former continent. Du Toit, Wegener's contemporary, was influenced by continental drift theory and improved on it greatly.

FORMATION OF THE CONTINENTS

Today continental drift theory is accepted widely, in large part owing to the development of plate tectonics, "the unifying theory in geology." We examine the evidence for continental drift, the arguments against it, and the eventual triumph of plate tectonics in the course of this essay. Before going on, however, let us consider briefly the now-accepted timeline of events described by Wegener and others. About 1,100 million years ago (earth scientists typically abbreviate this by using the notation 1,100 Ma), there was a supercontinent named Rodinia, which predated Pangaea. It split into Laurasia and Gondwanaland, which moved to the northern and southern extremes of the planet, respectively. Starting at about 514 Ma, Laurasia drifted southward until it crashed into Gondwanaland about 425 Ma. Pangaea, surrounded by a vast ocean called Panthalassa, formed approximately 356 Ma.

In the course of Pangaea's formation, what is now North America smashed into northwestern Africa, forming a vast mountain range. Traces of these mountains still can be found on a belt stretching from the southern United States to northern Europe, including the Appalachians.

As Pangaea drifted northward and smashed into the ocean floor of Panthalassa, it formed a series of mountain ranges from Alaska to southern South America, including the Rockies and Andes. By about 200 Ma, Pangaea began to break apart, forming a valley that became the Atlantic Ocean. But the separation of the continents was not a "neat" process: today a piece of Gondwanaland lies sunken beneath the eastern United States, far from the other landmasses to which it once was joined.

By about 152 Ma, in the late Jurassic period, the continents as we know them today began to take shape. By about 65 Ma, all the present continents and oceans had been formed for the most part, and India was drifting north, eventually smashing into southern Asia to shape the world's tallest mountains, the Himalayas, the Karakoram Range, and the Hindu Kush. This process is not finished, however, and geologists believe that some 250-300 million years from now, Pangaea will re-form.

EVIDENCE AND ARGUMENTS

As proof of his theory, Wegener cited a wide variety of examples, including the apparent fit between the coastlines of South America and western Africa as well as that of North America and northwestern Africa. He also noted the existence of rocks apparently gouged by glaciers in southern Africa, South America, and India, far from modern-day glacial activity.

Fossils in South America matched those in Africa and Australia, as Suess had observed. There were also signs that mountain ranges continued between continents—not only those apparently linking North America and Europe but also ranges that seemed to extend from

Argentina to South Africa and Australia. By measurements conducted over a period of years, Wegener even showed that Greenland was drifting slowly away from Europe, yet his theory met with scorn from the geoscience community of his day. If continents could plow through oceanic rock, some geologists maintained, then they would force up mountains so high that Earth would become imbalanced.

As for his claim that matching fossils in widely separated regions confirmed his theory of continental drift, geologists claimed that this could be explained by the existence of land bridges, now sunken, that once had linked those areas. The apparent fit between present-day landmasses could be explained away as coincidence or perhaps as evidence that Earth simply was expanding, with the continents moving away from one another as the planet grew.

PLATE TECTONICS

Though Wegener was right, as it turned out, his theory had one major shortcoming: it provided no explanation of exactly *how* continental drift had occurred. Even if geologists had accepted his claim that the continents are moving, it raised more questions than answers.

A continent is a very large thing simply to float away; even an aircraft carrier, which is many millions of times lighter, has to weigh less than the water it displaces, or it would sink like a stone. In any case, Wegener never claimed that continents floated. How, then, did they move?

The answer is plate tectonics, the name both of a theory and of a specialization of tectonics. As an area of study, plate tectonics deals with the large features of the lithosphere and the forces that fashion them. As a theory, it explains the processes that have shaped Earth in terms of plates (large movable segments of the lithosphere) and their movement. Plate tectonics theory brings together aspects of continental drift, seafloor spreading, seismic and volcanic activity, and the structures of Earth's crust to provide a unifying model of Earth's evolution.

THE PIECES COME TOGETHER

In 1962 the United States geologist Harry Hammond Hess introduced a new concept that would prove pivotal to the theory of plate tectonics: seafloor spreading, the idea that seafloors crack open along the crest of mid-ocean ridges and that new seafloor forms in those areas. According to Hess, a new floor forms when molten rock called magma rises up from the asthenosphere, a region of extremely high pressure underlying the lithosphere, where rocks are deformed by enormous stresses. The magma wells up through a crack in a ridge, runs down the sides, and solidifies to form a new floor.

Three years later, the Canadian geologist John Tuzo Wilson coined the term plates to describe the pieces that make up Earth's rigid surface. Separated either by the mid-ocean rifts identified earlier by Heezen or by mountain chains, the plates move with respect to one another. Wilson presented a model for their behaviour and established a global pattern of faults, a sort of map depicting the movable plates. The pieces of a new theory were forming, but as yet it had no name.

That name appeared in 1967, when D. P. Mackenzie of England and R. L. Parker of the United States introduced the term plate tectonics. They maintained that the surface of Earth is divided into six major as well as seven minor movable plates and compared the continents to enormous icebergs—much as Wegener had described them half a century earlier. Subsequent geologic research has indicated that there may be as many as nine major plates and as many as 12 minor ones.

To test these emerging ideas, the U.S. National Science Foundation authorized a research voyage by the vessel *Glomar Challenger* in 1968. On their first cruise, through the Gulf of Mexico and the Atlantic, the *Challenger*'s scientific team collected sediment, fossil, and crust samples that confirmed the basics of seafloor spreading theory.

These results led to new questions regarding the reactions between rocks and the heated water surrounding them, spawning new research and necessitating additional voyages. In the years that followed, the *Challenger* made more and more cruises, its scientific teams collecting a wealth of evidence for the emerging theory of plate tectonics.

REAL-LIFE APPLICATIONS

Early Evidence of Plate Tectonics

No single person has been as central to plate tectonics as Wegener was to continental drift or as the English naturalist Charles Darwin was to evolution. The roots of plate tectonics lie partly in the observations of Wegener and other proponents of continental drift as well as in several discoveries and observations that began to gather force in the third quarter of the twentieth century.

During World War II, submarine warfare necessitated the development of new navigational technology known as sonar. Sonar functions much like radar, but instead of using electromagnetic waves, it utilizes ultrasonic, or high-frequency, sound waves projected through water. Sonar made it feasible for geologists to study deep ocean basins after the war, making it possible for the first time in history to map and take samples from large areas beneath the seas. These findings raised many questions, particularly concerning the vast elevation differences beneath the seas.

Ewing and the Mountains Under the Ocean

One of the first earth scientists to notice the curious aspects of underwater geology was the American geologist William Maurice Ewing, who began his work long before the war.

He had gained his first experience in a very practical way during the 1920s, as a doctoral student putting himself through school. Working summers with oil exploration teams in the Gulf of Mexico off the coast of Texas had given him a basic understanding of the subject, and in the following decade he went to work exploring the structure of the Atlantic continental shelf and ocean basins.

His work there revealed extremely thick sediments covering what appeared to be high mountainous regions. These findings sharply contradicted earlier ideas about the ocean floor, which depicted it as a flat, featureless plain rather like the sandy-bottomed beaches found in resort areas. Instead, the topography at the bottom of the ocean turned out to be at least as diverse as that of the land above sea level.

Heezen and the Rift Valley

During the 1950s, a team led by another American geologist, Bruce Charles Heezen, worked on developing an overall picture of the ocean basin's topography.

Earlier work had identified a mountain range running the length of the Atlantic, but Heezen's team discovered a deep valley down the middle of the chain, running parallel to it. They described it as a rift valley, a long trough bounded by two or more faults, and compared it to a similar valley in eastern Africa.

Around the same time, a group of transatlantic telephone companies asked Heezen to locate areas of possible seismic or earthquake activity in the Atlantic. Phone company officials reasoned that if they could find the areas most likely to experience seismic activity, they could avoid placing their cables in those areas. As it turned out, earthquakes tended to occur in exactly the same region that Heezen and his team had identified as the rift valley.

The Plates and Their Interactions

The most significant plates that make up Earth's surface are as follows: Selected Major Plates

- North American (almost all of North America and Mexico,

along with Greenland and the northwestern quadrant of the Atlantic).
- South American (all of South America and the southwestern quadrant of the Atlantic).
- African (Africa, the southeastern Atlantic, and part of the Indian Ocean).
- Eurasian (Europe and Asia, excluding the Indian subcontinent, along with surrounding ocean areas).
- Indo-Australian (India, much of the Indian Ocean, Australia, and parts of the Indonesian archipelago and New Zealand).
- Antarctic (Antarctica and the Antarctic Ocean).

In addition to these plates, there are several plates that while they are designated as "major" are much smaller: the Philippine, Arabian, Caribbean, Nazca (off the west coast of South America), Cocos (off the west coast of Mexico), and Juan de Fuca (extreme western North America). Japan, one of the most earthquake-prone nations in the world, lies at the nexus of the Philippine, Eurasian, and Pacific plates.

Movement of the Plates

One of the key principles of geology is uniformitarianism: the idea that processes occurring now also occurred in the past. The reverse usually is also true; thus, as we have noted, the plates are still moving, just as they have done for millions of years. Thanks to satellite remote sensing, geologists are able to measure this rate of movement.

Not surprisingly, its pace befits the timescale of geologic, as opposed to human, processes: the fastest-moving plates are careening forward at a breathtaking speed of 4 in. (10 cm) per year. The ground beneath Americans' feet is drifting at the rate of 1.2 in. (3 cm) every year, which means that in a hundred years it will have shifted 10 ft. (3 m).

When Plates Interact

Plates interact by moving toward each other (convergence), away from each other (divergence), or past each other (transform motion). Convergence usually is associated with subduction, meaning that one plate is forced down into the mantle and eventually undergoes partial melting.

This typically occurs in the ocean, creating a depression known as an oceanic trench. Divergence results in the separation of plates and most often is associated either with seafloor spreading or the formation of rift valleys.

There are three types of plate margins, or boundaries between plates, depending on the two types of crusts that are interacting: oceanic with oceanic, continental with continental, or continental with oceanic. The rift valleys of the Atlantic are an example of an oceanic margin where divergence has occurred, while oceanic convergence is illustrated by a striking example in the Pacific.

There, subduction of the Philippine Plate by the Pacific Plate has created the Mariana Trench, which at 36,198 ft. (10,911 m) is the deepest depression on Earth.

When continental plates converge, neither plate subducts; rather, they struggle against each other like two warriors in a fight to the death, buckling, folding, and faulting to create huge mountain ranges. The convergence of the Indo-Australian and Eurasian plates has created the *highest* spots on Earth, in the Himalayas, where Mount Everest (on the Nepal-Indian border) rises to 29,028 ft. (8,848 m). Continental plates also may experience divergence, resulting in the formation of seas. An example is the Red Sea, formed by the divergence of the African and Arabian plates.

Given these facts about the interactions of oceanic and continental plates with each other, what occurs when continental plates meet oceanic ones is no surprise. In this situation, the oceanic plate meeting the continental plate is like a high-school football player squaring off against a National Football League pro tackle.

It is no match: the oceanic plate easily subducts. This leads to the formation of a chain of volcanoes along the continental crust, examples being the Cascade Range in the U.S. Pacific Northwest (Juan de Fuca and Pacific plates) or the Andes (South American and Nazca plates).

Transform margins may occur with any combination of oceanic or continental plates and result in the formation of faults and earthquake zones.

Where the North American Plate slides against the Pacific Plate along the California coast, it has formed the San Andreas Fault, the source of numerous earthquakes, such as the dramatic San Francisco quakes of 1906 and 1989 and the Los Angeles quake of 1994.

Paleomagnetism

As noted earlier, plate tectonics brings together numerous areas of study in the geologic sciences that developed independently but which came to be seen as having similar roots and explanations. Among these disciplines is paleomagnetism, an area of historical geology devoted to studying the direction and intensity of magnetic fields in the past, as discerned from the residual magnetization of rocks.

Earth has a complex magnetic field whose principal source appears to be the molten iron of the outer core. In fact, the entire planet is like a giant bar magnet, with a north pole and a south pole.

It is for this reason that the magnetized material in a compass points north; however, Earth's magnetic north pole is not the same as its geographic north pole. It so happens that magnetic north lies in more or less the same direction as geographic north, but as geologists in the mid-nineteenth century discovered, this has not always been the case.

In 1849 the French physicist Achilles Delesse observed that magnetic minerals tend to line up with the planet's magnetic field, pointing north as though they were compass needles. Nearly 60 years later, however, another French physicist, Bernard Brunhes, noted that in some rocks magnetic materials point *south*. This suggested one of two possibilities: either the planet's magnetic field had reversed itself over time, or the ground containing the magnetized rocks had moved. Both explanations must have seemed far-fetched at the time, but as it turned out, both are correct.

Earth's magnetic field has shifted, meaning that the magnetic north and south poles have changed places many times over the eons. In addition, the magnetic poles have wandered around the southern and northern portions of the globe: for instance, whereas magnetic north today lies in the frozen islands to the north of Canada, at about 300 Ma it was located in eastern Siberia.

The movement of magnetic rocks on Earth's surface, however, has turned out to be too great to be explained either by magnetic shifts or by regional wandering of the poles. This is where plate tectonics and paleomagnetism come together.

Confirmation of Plate Tectonic Theory

Rocks in Alaska have magnetic materials aligned in such a way that they once must have been at or near the equator. In addition, the orientation of magnetic materials on South America's east coast shows an affinity with that of similar materials on the west coast of Africa. In both cases, continental drift, with its driving mechanism of plate tectonics, seems the only reasonable explanation.

Thus, paleomagnetic studies have served to confirm the ideas of continental drift and plate tectonics, while research conducted at sea bolsters seafloor spreading theory. Using devices called magnetometers, geologists have found that the orientation of magnetic minerals on one side of a rift mirrors that of materials on the other side. This suggests that the new rock on either side of the rift was formed simultaneously, as seafloor spreading theory indicates.

Earthquakes and Volcanoes

Several findings relating to earthquakes and volcanic activity also can be explained by plate tectonics. If one follows news stories of earthquakes, one may begin to wonder why such places as California or Japan have so many quakes, whereas the northeastern United States or western Europe have so few. The fact is that earthquakes occur along belts, and the vast majority of these belts coincide with the boundaries between Earth's major tectonic plates.

The same is true of volcanoes, and it is no mistake that places famous for earthquakes—the Philippines, say, or Italy—often also are known for their volcanoes. Although they are located near the centre of the Pacific Plate, the islands of Hawaii are subject to plate movement, which has helped generate the volcanoes that gave those islands their origin.

At the southern end of the island chain, many volcanoes are still active, while those at the northern end tend to be dormant. The reason is that the Pacific Plate as a whole is moving northward over a stationary lava source in the mantle below Hawaii. The southern islands remain poised above that source, while the northern islands have moved away from it.

Chapter 13

Formation of Thunderstorms

People vary greatly in their reactions to thunderstorms. Some few take to feather beds in the fear that Thor will personally seek them out for punishment and destruction (though a feather bed actually avails against a lightning bolt about as effectively as a garden gate stops a railroad train).

Other people, perhaps equally foolish, rush to hilltops in the fond faith that Divine Providence will somehow ward any wandering electric bolt away from their exposed heads. Actually a thunderstorm is a beautiful and gigantic experiment in physical science, doubly interesting to those who know something of its causes and its mechanism, wildly beautiful to all men conscious of beauty in Nature.

On what earthly mountains does a sunset glow compare with its pure rose radiance on the white cloud-cliffs of a towering and detached thunderstorm? What ground outlook compares with the spectacle of a distant towering cumulus in the clear night sky, flickeringly lit and veined by an infinite and never-ceasing variety of soundless lightnings?

A great, wide-spreading, fast- traveling line squall, in its most violent and magnificent form, may be a towering range of such vapour-mountains, extending across the country for hundreds of miles and advancing in line like the fire-spitting front of some colossal and destructive army.

THOR'S APPROACH

Beautiful though thunderclouds are at a distance, either from the ground or from the air, on closer approach their still, turreted, snow-white serenity changes abruptly to a dark gray void of whirling winds, blinding firebolts, and torrential rain. If we (unwisely) fly into a thunderstorm in an airplane, this transition is very abrupt — one moment we are admiring the infinite overhang of a thousand cloud cliffs all gleaming with every gradation of sunlight and shadow; the next we are plunged into turbulent blue-gray obscurity, split by recurring yellowish flashes, deluged with rain, and disrupted by updrafts and air bumps of unbelievable ferocity. On the ground or the water, as the storm moves towards us at perhaps twenty or thirty miles an hour, the transition is slower and more obscure, yet spectacular and interesting enough.

It does not matter particularly whether we happen to be in the path of a single large summer 'air-mass' thunderstorm, or whether we are caught in the five-hundred-mile sweep of a spring or summer 'frontal' storm or line squall — the general effects will be the same, except that the line squall may be more violent, and its squall wind, temperature drop, and pressure rise will be more pronounced and permanent. In either case, reduced to typical form, the preceding summer weather is usually sultry and oppressive, both temperature and dewpoint being high.

The sky is generally clear, but somewhat obscured by haze and cirriform clouds out of the west. The wind is southerly, and perhaps light or occasionally lacking altogether with a single thunderstorm in the offing, but likely to be strong and gusty if a line squall (cold front) is bearing down from the west or northwest.

If detached cumulus clouds are watched closely, they can often be seen to grow vertically upwards, sometimes at surpris-ing speed. Towards the middle of the afternoon a wall of thunderheads may be glimpsed, beyond obscuring haze and lower clouds (and are often plainly to be seen from an airplane at high altitude), rising slowly out of the west. But the thunderheads themselves often stream out into cirrus-like nebulosity at higher levels, and this, combined with ground haze, may hide them from view.

In any case, the westering sun is soon blotted out, and the sky to west and northwest slowly darkens, at least in its lower regions, to an ominous 'blue-black colour. Thunder begins to be heard perhaps half an hour to an hour before the storm strikes, and becomes louder and more frequent by the minute. The wind, which has perhaps died down altogether, rises and begins to blow directly towards the storm. Perhaps fifteen minutes before the squall strikes, the 'squall head' or roll cloud, whitish against the black clouds above and behind it, may be seen advancing in the shape of a long and irregular roll across the front of the storm — and this roll sometimes visibly revolves like a sort of horizontal tornado.

Perhaps fifteen minutes or more before the storm strikes, also, distinct lightning bolts come into view, most of them striking directly downward into the ground from the bottom and front of the black clouds just behind the roll cloud. We can check the narrowing distance away of these fiery ground-strikes, of course, by noting the flash-thunder time interval with a stop watch — the lag between sight and sound (as sound travels about eleven hundred feet per second in air) is about five seconds per mile. After the lightning bolts approach within a mile or two, it is only a moment (in a severe storm) before they are striking all around us, and in this 'dry' front of the storm, before the rain begins, the lightning danger is greatest.

A few minutes after the roll cloud passes overhead, the squall wind sweeps out of the west or northwest, usually at something like thirty or forty miles an hour, but occasionally as a flattening blast of sixty or more. Along with the squall, or shortly after its onset, comes a wall of spattering, large-drop rain that may descend in torrents, and may be accompanied by lashing hailstones.

In a detached 'air-mass' storm the squall wind usually begins to moderate after a few minutes, and may have died away altogether in a half-hour or so. The rain is less heavy, and the lightning less frequent and less intense, in the rearward and less active parts of the storm. But after a line squall the high westerly wind may continue or even increase.

THUNDERSTORMS OCCUR

Thunderstorms occur most frequently in the warm, moist, and unstable air masses within the tropics, where (in the rainy season) they are a daily afternoon occurrence, and sometimes so regular that they serve as a convenient time indicator for human affairs. In arctic regions, where the air is cold and comparatively dry, thunderstorms may not occur oftener than once in several years. Temperate regions experience their share of storms in the late spring and summer. No part of the United States is free from thunderstorm activity, but different regions vary widely in thunderstorm frequency.

In winter the centre of electrical activity is over the lower Mississippi Valley; in summer there are two far more active centers, one over Florida and the eastern Gulf coast, the other over northern New Mexico. But these centers are only relative, of course, for many parts of the nation see nearly as many storms. And in some regions, such as the Central, North Central, and East Central States, thunderstorms are likely to be mostly of the frontal type, and hence more severe and destructive.

Thunderstorms are classed, according to their apparent severity, as 'mild,' involving few if any ground-strikes and relatively light winds; 'moderate,' involving fairly frequent groundstrikes, moderate to heavy rain, and moderate to strong winds; and 'severe,' attended by nearly incessant lightning, torrential rain, and severe squalls. Temperate-zone thunderstorms are also classed, according to the mechanism of their formation, as 'airmass' and 'frontal.' The older names for these same classes of storms, before air-mass-and-front analysis came into wide use, were 'convective' and 'cyclonic.'

Air-mass thunderstorms occur entirely within one air mass, and without aid from any frontal action between air masses. They arise where the characteristics of the air mass, aided by wind and topography, favour them. They are most frequent in the United States in the warm, moist, and conditionally unstable air masses of tropical maritime origin. But moist polar air masses advancing southward rapidly, and hence remaining relatively cold in their upper levels, can occasionally be unstable enough to produce scattered air-mass thunderstorms.

The atmospheric instability that produces air-mass thunderstorms and keeps them active for hours over land areas in daytime is usually 'triggered off' by intense solar heating at the surface, which steepens the vertical temperature gradient; or perhaps by the importation of cold air aloft; or by the rising currents produced by upward deflection of horizontal winds (together with perpendicular solar heating) on the slopes of hills or mountains. Thus air-mass storms are frequent along the Gulf coast, where the sea air strikes the hot and slightly rising land; and thunderstorms of any type are more frequent over mountains than over flat country or at sea.

Frontal thunderstorms may occur along a warm front, where even gradual lifting on the warm-front slope is sufficient to trigger off the instability of very warm and very moist tropical air. They may also occur along the wave-like oscillations of a more or less stationary front that is, for the moment, alternately warm and cold.

They may even arise, rather mysteriously for unaccustomed forecasters, out of upper fronts that give little or no other indication of their motivating presence. But the great majority of frontal thunderstorms occur either directly in front of, or actually ranged along, a vigorous cold front marked on the weather map by large temperature contrasts, a deep pressure trough, and strong shifting winds (that is to say, by a fairly strong pressure field having isobars sharply bent into V-shape where they cross the front). This whole formation is, of course, nothing more or less than a line squall.

Thunderstorms of all types are most frequent, on land, during the middle of the afternoon — that is, during the hours when solar heating at the surface reaches its maximum and the lapse rate (vertical temperature gradient) is consequently steepest. As the average life of a thunderstorm is six or seven hours, however, thunderstorms formed in the afternoon often persist into the early evening. Over the ocean, where the equable water warms up by only a degree or so during the day, and cools off by even less at night, the thunderstorm maximum comes during the late night and early morning hours, when radiational or airmass-transport cooling aloft is greatest, and, again, when the total lapse rate is steepest.

As the mechanics of frontal thunderstorms depend not so much on surface heating or upper cooling as on the violent upthrusting of warm and moist air masses by an undercutting wedge of cold air, frontal storms may occur at any time of the day or night. But even the frontal storms are most likely, and most severe, during the middle of the afternoon on land or during the early morning hours at sea.

Plenty of active summer cold fronts, for example, do nothing much over land during the morning hours, yet develop into roaring line squalls by mid-afternoon.

FORMATION OF A THUNDERSTORM

What are the aerological conditions that favour the formation of a thunderstorm? First, there is the all-important factor of atmospheric instability, evidenced by a steep or steepening lapse rate (-3° F/1000 ft or steeper, or about -6° C/km or steeper). This steep temperature gradient is perhaps being increased by air-mass transfer below or aloft — lower warm winds from the south and southeast and/or upper cold winds from the north and northwest — and above all, by excessive solar heating at the surface.

In addition to the steep temperature gradient, high relative humidity at all levels (meaning that clouds will be formed by very little rise of air) also favors the formation of a storm. Finally the 'ice level,' or altitude of freezing temperature (+32° F. or 0° C.), should be located at considerable elevation, as it always is over the tropics. These upper-air characteristics, of course, can be fully known in advance only by means of airplane or balloon soundings, though they can sometimes be sensed by the rapid (and occasionally prodigious) vertical growth — the spectacular upward mushrooming — of detached cumulus clouds during the late morning.

There are also some surface thunderstorm portents not without value — high temperatures, say 80, 90, or 100° F. by day and 70° or 80° by night; high surface specific humidity as indicated by high dewpoint, say more than 6o° or 70°; and steadily decreasing surface pressure during the late morning and early after. noon. Another useful early-afternoon indication is increasing 'crash' static from a radio set — actually radio waves from lightning flashes beyond the horizon.

Finally there are the definite evidences of frontal activity and advance towards one's position to be obtained from some sort of weather map — even a simple regional map that anyone can draw from airways radio weather reports in a few minutes' time.

One or more of the indications just listed can of course be present without resulting in thunderstorms, and thunderstorms can occur when one or more of them are absent; but in general the presence of most of them makes thunderstorms likely, and the agreement of all of them makes thunderstorms certain.

Given these favorable conditions, how does an air mass thunderstorm actually form and grow? The beginning of a thunderstorm is similar to the formation of an ordinary cumulus cloud or an ordinary shower. But in the case of the thunderstorm the air is more moist and more unstable, and this moist instability extends to greater heights. Hence the updrafts are more violent, and the cloud building is more rapid and extensive.

As in ordinary cumulus-building, the rising currents of a thunderstorm cause the warm, moist air to cool by expansion to its dewpoint, forming minute liquid droplets (on commonly present condensation nuclei), perhaps 1/2500 inch in diameter, which droplets appear as the white or gray opacity of the cloud; and this condensation of water vapour, in a vicious cycle, releases latent heat which helps to drive and intensify the rising currents, which condense further moisture, and so on. The result is that under favorable conditions the thunderstorm grows fast and far — perhaps up to altitudes of six or seven miles.

Just as in the formation of all rain, so, while the minute cloud droplets of a thundercloud are carried up in the rising currents, there is some jostling between them, which jostling tends to coalesce many of them into much larger drops. These larger drops, if the updraft is not too strong, will begin to fall through the cloud as rain. If the upsurging cloud droplets reach a temperature as low as +5° F. (-15° C.), which they soon do in a

by high dewpoint, say more than 60° or 70°; and steadily decreasing surface pressure during the late morning and early after. noon. Another useful early-afternoon indication is increasing 'crash' static from a radio set — actually radio waves from lightning flashes beyond the horizon. Finally there are the definite evidences of frontal activity and advance towards one's position to be obtained from some sort of weather map — even a simple regional map that anyone can draw from airways radio weather reports in a few minutes' time.

One or more of the indications just listed can of course be present without resulting in thunderstorms, and thunderstorms can occur when one or more of them are absent; but in general the presence of most of them makes thunderstorms likely, and the agreement of all of them makes thunderstorms certain.

Given these favorable conditions, how does an air mass thunderstorm actually form and grow? The beginning of a thunderstorm is similar to the formation of an ordinary cumulus cloud or an ordinary shower. But in the case of the thunderstorm the air is more moist and more unstable, and this moist instability extends to greater heights. Hence the updrafts are more violent, and the cloud building is more rapid and extensive.

As in ordinary cumulus-building, the rising currents of a thunderstorm cause the warm, moist air to cool by expansion to its dewpoint, forming minute liquid droplets (on commonly present condensation nuclei), perhaps 1/2500 inch in diameter, which droplets appear as the white or gray opacity of the cloud; and this condensation of water vapour, in a vicious cycle, releases latent heat which helps to drive and intensify the rising currents, which condense further moisture, and so on. The result is that under favorable conditions the thunderstorm grows fast and far — perhaps up to altitudes of six or seven miles.

Just as in the formation of all rain, so, while the minute cloud droplets of a thundercloud are carried up in the rising currents, there is some jostling between them, which jostling tends to coalesce many of them into much larger drops.

These larger drops, if the updraft is not too strong, will begin to fall through the cloud as rain. If the upsurging cloud droplets reach a temperature as low as +5° F. (-15° C.), which they soon do in a towering cloud (at a height of four or five miles in temperate zone summer), some of them are jostled into ice particles; and ice particles are perhaps also subliming directly out of water vapour (onto sublimation nuclei already present at these cold heights), if the temperature and humidity conditions are favorable.

In a fully developed thunderstorm the ice-particle cloud is often visible as a cirrus-like, anvil-shaped top or cap that blows out somewhat ahead of the main cumulus turret on the upper winds. The myriads of ice particles, falling slowly and growing into larger crystals, eventually produce rain and other forms of precipitation.

In this fashion the production of rain continues until large drops perhaps a quarter of an inch in diameter are falling through the cloud; and the rising currents build up until, in the forward and most active part of the storm, they exceed twenty miles an hour.

This updraft will just support a three-eighths-inch raindrop; and no raindrop larger than three-eighths inch can exist in nature, for it would be instantly blown into smaller drops by the wind of its fall. Thus no rain can fall through the most active updrafts of the thundercloud, and the active front of the storm is marked by an abrupt stoppage of large-drop rain falling from above, by the deflection of this rain upward towards the rear of the storm, and by the breaking up of large drops into smaller drops.

From the breaking up of these water drops, which can, of course, recombine above and descend to be broken again, come the electric charges that produce the lightning. At the instant of spattering, the to-be-broken drops are charged with positive electricity, and the air surrounding them is negatively charged.

This negatively charged air is carried away into upper and rearward portions of the thundercloud, but the positively charged raindrops, breaking and recombining, tend to accumulate in the most active (forward and lower) portion of the storm, which hence becomes a region of high positive electrical charge.

From this positively charged region, usually just above and behind the roll cloud or 'squall head,' frequent heavy thunderbolts strike downward and forward directly into the ground (whose charge is neutral or somewhat negative), while other longer but less intense flashes strike upward and backward to lose themselves in the upper immensities of negatively charged cloud.

Towards the rear of the storm, however, where the earth is positive with respect to the negative clouds above, less intense bolts strike upward from the ground into the clouds.

Lightning may be only one feature of a great storm perhaps marked also by cloudbursts, ice showers, and destructive blasts of wind, but it is spectacular enough, and under some conditions hazardous enough, to warrant examination in some detail before considering the other aspects of the storm.

THOR'S HAMMER

The potentials set up by the cloud-formed electric charges of a thunderstorm are enormous — in the neighborhood of one hundred million volts — and they discharge through dry air, ordinarily considered one of the best insulators, to usual distances of half a mile, or even occasionally to a mile or more. The lightning bolt itself, though similar in many ways to a small spark in the laboratory, yet shows additional and peculiar characteristics introduced by its enormous voltage and the great length and diversity of its path.

There is nothing oscillatory or alternating about the lightning bolt — it is one or more surges of current in the same direction. Moreover the bolt does not jump instantly from a positive cloud, say, to the negative earth — it appears to grow or burrow through the air in a succession of electronic darts, building its own ionized (or electrically conducting) path as it goes.

Ordinarily this growth is frightfully rapid, and takes place in perhaps one millionth of a second. But occasionally it is slow enough to be seen by the naked eye in the form of 'rocket lightning,' in which a growing and perhaps upward-directed bolt burrows visibly at no greater speed than an ordinary rocket.

Some of the veined lightning in a detached thunderhead, viewed from several miles away on a clear night, can often be seen to grow or burrow in this fashion progressively along the bulges of cloud. Usually lightning strikes from thundercloud to ground, from ground to cloud — or, even more commonly, between clouds. But on rare occasions it flickers or glows out of a clear sky, as charged masses of air pass near each other or the ground.

The amperage or rate of electric current flow in a thunderbolt is likewise enormous, compared with most man-made currents. Its value ranges all the way from ten thousand amperes in weaker flashes up to perhaps two hundred thousand amperes in larger bolts. One hears of people being struck by lightning and surviving the ordeal, but these fortunates are probably grazed by side flashes or forked branches while the main bolt goes elsewhere.

Few if any animals, human or otherwise, have ever been hit by a heavy main bolt and lived to tell the tale. For like its voltage and amperage, lightning power is unbelievably large — perhaps ten million million kilowatts or thirteen trillion horsepower! But as an average single flash lasts only three or four millionths of a second, the total amount of electricity (or electrical energy) involved is not very large — perhaps three thousand kilowatt hours or so, obtainable commercially within more usable limits of current and voltage for seventy to a hundred dollars.

Lightning punishment at one point is by no means limited to a single bolt, however, and the old adage that 'lightning never strikes twice in the same place' is about as untrue a statement as anyone could conveniently concoct. Many lightning flashes are of the 'multiple' variety; the bolt not only strikes twice, but may strike repeatedly — five times, ten times, or even forty times in the course of a second or two. This multiple lightning usually appears to the eye as a single, flickering, long-lived bolt, and it is the kind most likely to start fires in any dry and combustible material along its path. The brilliant flash that accompanies a lightning discharge, formerly thought to be caused by sudden heating of air, is now known to result from ionization of the air by the terrifically strong electric field along the path, causing the very atoms of dense air to vibrate at such high frequencies that they radiate visible light in the same way that rarified neon in a glass tube glows under moderate electrification.

The lightning flash is thus kindred (though once or twice removed) to the flickering aurora of polar skies, and to the glowing message of an advertising sign.

Lightning is usually white in colour, combining the spectrum of nitrogen with that of oxygen. But it may appear bluish in contrast to yellowish artificial lights, and may appear yellowish if viewed through haze, fog and clouds. Occasionally a reddish or pinkish flash occurs, when air rich in water vapour is ionized so as to give the spectrum of hydrogen.

Thunder takes many audible forms, but they all originate in one underlying cause — the sudden increase of air pressure along the path of the bolt (for any sound wave is only a sudden increase in air pressure) resulting from the heating, molecular dissociation, and ionization of air by the bolt.

Earlier vague assertions about 'air collapsing to fill up the hole hollowed by the bolt,' and such notions, are now in the discard. If lightning strikes close enough — within a few feet of us — the sound is simply one sharp and terrific 'bang!' Like that of a large cannon fired in the same place; for the ear is too paralyzed by this first and nearest blast to notice the subtler and later gradations of reverberating thunder arriving from more distant portions of the bolt.

But if the bolt is farther away, and particularly if it is a long one, the thunder arriving successively (at eleven hundred feet per second) from its more and more distant portions has the sound of a long roll, which wavers considerably even as the characteristics of the bolt, and its direction with respect to the hearer, vary greatly along its path.

Under certain conditions of air stratification, and often in the presence of steep mountains and cliffs that also serve to reflect and intensify sound, the roll of the thunder echoes back and forth; these overlapping echoes, together with simultaneous sound reinforcements from equidistant portions of the path, may build up into great booming crescendos of noise. Thunder is seldom heard over horizontal distances of more than fifteen miles; and if the flash occurs at altitudes of more than two miles, it may not reach the ground at all in audible form.

Several kinds of lightning have been observed, of which by far the most common is 'streak' lightning — the ordinary bolt. A streak of lightning is not a zigzag, as commonly drawn, but sinuous in shape, like a river in broken country. The streak may be single, but more commonly, part of it splits off into smaller downward branches that may or may not reach the ground.

Occasionally a bolt splits into 'forked' lightning — two or more main branches that each strike home. Multiple streak lightning, drifted sideways by a strong wind, sometimes appears as 'ribbon' lightning. And if a single streak follows a very sinuous and varied path, it may present the illusion of glowing beads where it is either more intense, or else where part of the flash is seen end-on.

Another distinct variety, very rare but very dangerous, is 'ball' lightning. Scientists long doubted altogether the existence of this variety, but so many accounts of its appearance were authenticated that it has come to be believed in, and more or less explained. It appears in the form of a luminous ball, perhaps the size of a clenched fist, which either falls through the air, floats horizontally, or rolls along some solid object such as the ground.

Visible only for a second or two, it disappears either quietly, or with a terrific explosion. According to present theory, ball lightning is nothing more than a creeping corona discharge along a relatively intense portion of the electric field — or, one might say, a 'stalled thunderbolt,' similar in kind to rocket lightning but much slower. If the electric charge behind it dissipates, the ball fades away also; but if the original charge increases enough to break down the whole path with a large thunderbolt, as it well may — then look out!

Another variety of Nature's electric illumination is 'sheet' lightning. This is not to be confused with so-called "heat' lightning — the diffuse and silent reflection of distant electric flashes hidden either by clouds or by the horizon. True sheet lightning occurs as a momentary glow extending through fairly large masses of air in the cloud region (or troposphere), and is nearly always white in colour. It is probably an ionization effect similar in principle to the aurora, though the sheet-lightning glow occurs on a much smaller scale than the aurora, and at much lower levels in the atmospheric ocean.

Another and somewhat similar variety of thunderstorm light is ' Saint Elmo's fire,' a bluish, brush-like discharge that sometimes appears on rooftrees, tall steeples and chimneys, the mastheads of a ship, or any other high, sharp-pointed object during a thunderstorm. It is merely the well-known corona discharge, reproducible at will in the laboratory or on a high-voltage transmission line.

Occurring in nature, it indicates intensification of the electric field, and may indicate the imminence of a thunderbolt. Such a corona discharge, of course, presents a very real fire hazard — particularly in the case of a hydrogen-filled balloon or airship. On May 6, 1937, during a great line squall over New Jersey, the German airship *Hindenburg* was valving explosive hydrogen in preparation for a landing at Lakehurst. Saint Elmo's fire ignited the escaping hydrogen-air mixture, and the great airship crashed in flames, killing thirty-six of its transatlantic passengers.

This incident marked the tragic end of hydrogen-filled airship activity; and as the United States Government wisely declined to release helium to a nation arming to the teeth for a treacherous and unprovoked attack on its neighbors, it marked the effective end of all German airship activity for some time.

LIGHTNING HAZARDS

Just as lightning bolts themselves exhibit almost endless varieties of form and intensity, so do their earthly effects vary greatly, not only with the bolts' severity, but even more with the things and substances which a ground-strike encounters on its path to earth. Metallic conductors of sufficient size — say a bar of copper or other metal the size of a pencil — pass the bolt without harm to themselves or anything else.

But if the conductor does not offer the shortest and easiest path to earth (from the viewpoint of the hop-along bolt itself), it may jump to a more tempting and destructive path; and even if the main bolt follows the conductor, side flashes may jump to neighboring conductors, be they metal tanks or pipes, pools of water, or human beings.

Small, hollow conductors may be crushed by the intense electrostatic and electromagnetic fields accompanying lightning passage. Irregularly shaped, loose, or insecure rods or wires are likely to be moved, bent, burned, or melted; and insufficiently large conductors, such as a radio antenna, are invariably melted or vaporized instantly by a direct stroke.

Non-conductors such as wood and stone fare much worse, at least during the 'dry' part of the storm before they are wetted over with a film of partly conducting water. Dry trees and poles are often split and splintered throughout their length by the explosive action of their moisture content suddenly converted into steam, and the pieces are blown forcibly outward in all directions. Dry or partly moist bricks and mortar may be violently exploded, and very rarely a small and flimsily built house is wrecked completely, as if by an earthquake.

Some of lightning's worst destructive effects are seen where large conducting masses are separated by non-conducting ones -such as a metal roof, ungrounded, on top of a wooden house. Here the conductor seems to collect and intensify the bolt, which wreaks havoc on the non-conducting structure beneath. No ordinary insulator such as a glass plate, of course, offers any effective resistance to a lightning bolt. If an insulating plate stands in the way and cannot easily be got around, the bolt simply blasts or burns a hole right through it.

The best possible form of lightning protection for any object, animate or inanimate, is a 'Faraday cage' — in complete form a grounded shell of metal not less than one twentieth inch thick, completely surrounding the object in question and several feet away from it on all sides. But as such a metal box would be very expensive and inconvenient, it can be approximated, with nearly as good practical results, by a skeletonized cage of interconnected conductors — each conductor being at least one fourth inch thick, if of copper; or one third inch thick, if of iron.

Of such general form are the lightning-rod systems now in use on thousands of isolated houses and barns in thunderstorm areas — systems that prove their usefulness by greatly reducing fires and other lightning hazards. In general, these protective rods should extend at least a foot or two above all the higher projecting parts of a building; they should extend downward as directly as possible to thoroughly good ground connections; and they should be bonded to each other and to any other large metallic masses in the building, which latter should be independently grounded.

Any high metallic conductor, such as a steel tower or a steel-frame city skyscraper, usually confers about it a 'cone of protection' having a radius equal to three or four times the height of the tower; occasionally, however, this cone of protection is violated by an erratic bolt. But for all the unpleasantness of lightning in the open — the vivid flashes of uncomfortably close ground-strikes and the deafening crash of near-by thunder — the actual hazard is really very small.

Out of all the millions of people exposed to lightning in the United States, only about five hundred are killed by it each year. Under the intense part of a very severe storm, however, and in various unwise locations beneath any storm, the risk may notably increase. The Bareau of Standards has issued some rules of wise personal conduct for those caught out in the open by storms, which are as follows:

- Do not go out-of-doors or remain out during thunderstorms unless it is necessary. Stay inside of a building where it is dry, preferably away from fireplaces, stoves, and other metal objects.

- If there is any choice of shelter, choose in the following order:
  - Large metal or metal-frame buildings.
  - Dwellings or other buildings which are protected against lightning.
  - Large unprotected buildings.
  - Small unprotected buildings.
  - If remaining out-of-doors is unavoidable, keep away from:
  - Small sheds and shelters if in an exposed location.
  - Isolated trees.
  - Wire fences.
  - Hilltops and wide open spaces.
- Seek shelter in a cave, a depression in the ground, a deep valley or canyon in dense woods, or a grove of trees.

Their cumulus clouds are merely prosaic banks of condensers, but their lightning is real and convincing, as anyone can testify who has seen one of their five million-volt flashes strike downward in the dark through fifteen feet of air to shatter a three-inch fence post with the bang of a small cannon.

Safety of People in Ordinary Wood-Frame Houses

Much will depend upon the construction of the house and what is in it, and particularly the amount of metal in the walls, whether in the form of gas, water, or vent pipes, how the electric wiring is arranged — whether in rigid conduits or in BX, or knob-and-tube open wiring. Quite obviously, the only absolutely safe place to be is inside of a completely enclosed metal structure, which may or may not be grounded.

The degree of safety enjoyed by the occupants of an ordinary wood-frame house will depend upon the degree of approach to this completely shielded structure. The wave front of current discharge in a direct stroke is steep enough, and the magnitude great enough, so that in many cases fairly high potentials will be built up in conductors of thirty- to forty-foot lengths — so that side-flashes may occur to other metal structures which may be connected to the same ground but at some distance away.

This was particularly exemplified in a case where a man had a lightning rod system properly installed, as far as the lightning rods were concerned, and yet it flashed through about six inches of wood and air to a conduit located in the peak of the attic roof underneath the lightning rod system. Of course if these two had been tied together at this point, the fire which resulted would have been avoided.

So far as my experience goes, the greatest danger from lightning in the ordinary wood-frame house is from fire, and of course this is particularly so in the rural communities where fire-fighting apparatus is not readily available. In cities the damage done to buildings as a result of lightning is not as a rule serious, although cases are on record where explosions have occurred during lightning storms, but apparently these explosions were due to the existence of gas or other explosive materials enclosed within the house at the time it was struck.

Wood-Frame Houses Protected by Lightning Rods.

Insurance records for the State of Iowa indicate a very substantial improvement, which is also substantiated by those from the Province of Ontario, as reported by the Fire Marshal of Ontario. These results have indicated that lightning protection is better than ninety per cent efficient, from the point of view of preventing fires in protected buildings.

Of course not all of the lightning protection systems were properly installed. Many were, no doubt, inadequately grounded, which accounts for the imperfect record to some degree. At the same time, we know today that lightning discharges do not always strike the highest object, and that now and then the cone of protection, which laboratory tests have demonstrated exist about a lightning rod, is violated.

I have, for instance, a photograph showing lightning striking the top of the Empire State Building, in which the discharge failed to contact the very top of the building but struck about forty feet down. It is apparent that lightning discharges take a path which might well be called the best path, from the point of view of the discharge itself.

Safety of People in Closed car with steel Turret top

Following the principle of the Faraday cage, there is no reason why people inside of a car with a steel top should not be very safe from lightning discharges, provided that they are not in the process of getting in or getting out of the car, or do not have their heads or arms out of windows, or are not standing on the ground changing a tire, and assuming that the gas tank is not struck and punctured, which is relatively unlikely. I believe the best advice to a man riding in an automobile during a lightning storm is to keep going and not to get out of the car, and the chances are that he will be perfectly safe.

Of course, if power wires fall on the car, which has happened during violent storms, he is wise to remain in the car, and not attempt to get out, since this procedure has in the past resulted in serious accidents.

Safety of People in Closed Wooden Yacht

As a general thing, lightning is more likely to strike a wet mast on the water, if it is striking round about, than a similar object on land, because of the relative smoothness of the water and the relative height of the projecting mast.

It is probable that suitable protection could be provided to the boat and its occupants by connecting a conductor following down the mast to the metal keel of the boat. In case it is an engine-driven boat, care should be taken to be sure that all the metal objects within the boat which are not connected together are electrically bonded together, and connected to some outside metallic conductor, such as the keel, which furnishes good contact with the water.

I don't know of any definite rule which has been set up concerning the protection of boats, but it seems to me that the general principle of the Faraday cage, or its equivalent, is one which can be applied to boats, coupled with the idea of the lightning rod and its usual cone of protection, and means for discharging the current of the lightning discharge itself into the water, without having either the boat or its occupants become a part of the circuit.

It might be well at this point to inject an observation concerning people in swimming, to the effect that, in my estimation, the danger to the swimmer is not so much that he may be struck by lightning, as that a small current flowing through the water, even at considerable distance from the stroke itself, may be sufficient to render him unconscious, and he dies by drowning, rather than by electrical shock.

Safety of People in an Airplane

Once more the question is one of preventing the passenger or pilot from becoming a carrier of current of the lightning discharge by being in series with it.

The only known way to prevent this is either to keep the lightning current from entering the plane by means of an all-metal plane, or its equivalent protection by some means of shunting the discharge which gets into the plane from the body of the occupants, and means also of preventing the lightning discharge from interfering with the mechanical and electrical operation of the plane.

The best method of doing this is probably the all-metal plane, which brings us to the Faraday cage, where if the joints in the plane are thoroughly bonded together, and the metal parts so connected that the measuring instruments and other electrical circuits of the plane are not likely to become a part of the lightning path, the possibility of damage is relatively small.

The occupants of the plane ought to be perfectly safe provided they are entirely surrounded by metal having a reasonable current carrying capacity....

From the point of view of fire, the multiple stroke is undoubtedly far more serious than the single stroke, and I believe I am quite safe in stating, that the single stroke of large magnitude is probably responsible for the explosive effects frequently observed, whereas the multiple stroke, even though of many relatively small currents, may be responsible not only for explosive effects but for fires also, provided inflammable material is present. As many as forty strokes have been observed within one second of time, and data which we have been taking during the past two years seem to indicate by far the great majority of lightning strokes are multiple....

DOWNBLAST AND DELUGE

A typical thunderstorm forms out of nothing into a towering Everest of threatening cloud, how the terrific electric potentials of the lightning arise out of a mere spattering of raindrops. But what about the lashing hail, the scouring squall winds, and the other thunderstorm accompaniments that often prove more dangerous and destructive than the lightning itself?

We had earlier in mind the picture of a towering cumulus cloud marked by fast-rising air currents and fast-falling raindrops of near-maximum size, whose spattering in the zone of strongest updrafts produced positive electric charges large enough to forge the thunderbolts.

As most of the warm, moist air contributing its latent heat to these rising currents is coming from in front of the storm, the most violent updrafts are usually in the front part of the main thundercloud; and they tend to slope somewhat towards the rear of the storm, carrying the rain rearward also. Just behind the main updrafts, where their speed has fallen off to less than twenty miles per hour, will be the region of heaviest rainfall.

Much of this heavy rain, coming from upper and colder regions of the thundercloud, is cold; some of it may even be frozen. This cold deluge cools the air as it falls, partly by its own coldness, but even more by its partial evaporation. The result is that air in the heavy rainfall region is soon chilled well below the surrounding air, hence becomes denser than the surrounding air, hence begins to sink, and soon develops into a great downdraft of cold air perhaps nearly as violent as the updraft of warm air in front of it.

This cold downdraft reaches the ground with the forward speed of the whole storm at upper levels (which may be twenty or thirty miles per hour or more), and at the ground level it sweeps forward as a violent outrushing squall wind, blowing in heavy gusts directly out of the storm.

But the squall wind may far exceed the speed of the storm itself. It is not impossible for a storm moving at thirty miles an hour, say, to produce a surface squall which (however shortlived) may blow at sixty or eighty miles per hour and sweep all before it. These excessive squall winds are perhaps partly explained by the crowding of the downdraft against the ground (which also contributes to the sharp barometer rise noted with the squall).

But the chief reason for their ferocity is the action of a horizontal vortex. With a great updraft in one part of the cloud and a great downdraft behind it, all sorts of eddies and minor whirlwinds are bound to exist in thunderclouds, as airmen who have endured the blind bumpiness inside one can fervently testify.

Most of these vortices, being well up in the cloud, are invisible (except on the gyrating instrument dials of an unhappy airplane); but one of them, being formed just at the cloud base, where the relative humidity is such that the slightest drop in pressure condenses moisture, may plainly appear as the long, light-colored roll cloud or 'squall head' arching horizontally across the front of the storm.

As the roll-cloud vortex revolves, it moves across land or water like the revolving brush of a gigantic carpet-sweeper, piling its speed of rotation on top of its speed of horizontal motion — all to the consternation of squall-blown humans in one case, or insects swept out of a carpet in the other. This horizontal vortex of the roll cloud is, in fact, a sort of horizontal tornado; and under certain conditions, parts of it can turn down to the ground and become real tornadoes. In any case, it is something for an aviator to avoid at all costs.

This whole thunderstorm wind-mechanism picture, has, of course, been snapped at a single and very typical instant. Actually, thunderstorms and line squalls vary greatly in size, form, and intensity. Some thunderstorms have no squalls; some squalls have no thunder; and any one storm is constantly changing, from moment to moment, its form and intensity. Updrafts and down drafts alike are constantly changing their speed and their position in the storm.

Thundershowers are usually of the brief, heavy type, and may reach cloudburst intensity. Raindrops in the active parts of the storm are much larger than ordinary raindrops, being mostly of near-maximum size; and they are crowded together, in certain places, by the interplay of violent and ever-varying air currents. It is not at all uncommon for these thunderstorm deluges to fall at the rate of three inches or more an hour, flooding low places with a few feet of suddenly rising water.

On occasion, the rate of water-fall for a brief interval has been known to run as high as one inch per minute! To a ground observer such a cloudburst seems to be falling in sheets of solid water, as if the heavens had indeed opened; to an aviator trying to fly through the downpour, it seems almost like boring through the depths of the ocean, and airplanes have actually been brought down by running unsuspecting into a heavy cloudburst at night.

On the ground a cloudburst is most destructive in normally dry desert and mountain regions — among steep converging slopes bare of brush or timber — where any dry ravine may within a few minutes become a raging torrent, and walls of tumultuous water ten or twenty feet high may roar along the rock-walled bottoms of steep canyons.

Nor is torrential rain the last of possible thunderstorm visitations on men caught in or below the storm. On occasion, thousands of tons of ice may fall from the skies in the destructive form of hail. The upper parts of a towering thundercloud harbor many slowly falling ice particles. Under favorable conditions of temperature and humidity, some of these may grow into sphero-crystals or snow balls (*graupel*) about the size of small peas.

Alternatively, or in addition, raindrops carried by the storm's great updrafts into temperatures well below freezing, and sufficiently jostled by turbulence, may be frozen into ice balls (sleet) of similar size. When the updraft becomes less violent, or when these pellets reach the edge of it, they fall groundward again, perhaps gathering a coating of snow above the freezing level, and a further coating of slush or water below it.

Soon, however, the growing ice pellet may enter new violent updrafts, which carry it upward again into the arctic zone, where its acquisitions of snow, slush, and water are frozen into a new concentric shell of solid ice, increasing its diameter. In like manner the ice pellet may fall again through several thousand feet, be carried upward again through another great rise, and repeat the whole down-up cycle ten or twenty times, growing the while into an ice ball of considerable size.

The largest hailstones ever authentically observed, which fell on some Nebraska farms in 1928, were the size of small grapefruit. One of these stones weighed one and one half pounds and was about six inches in diameter! Three-inch hailstones, the size of ordinary oranges, probably shower down somewhere in the United States a few times every summer.

Stones of golf-ball size are not at all uncommon. In 1936, nineteen Bushmen in South Africa were reported to have been killed by a fall of oversized hail that covered their bodies to a depth of three feet. Hail of some sort accompanies a large proportion of temperate.

One thunderstorms, but the hail zone of any one storm is usually not more than a few miles wide. But in tropical thunderstorms hail is very rare, however heavy the lightning and rain, because rain-making portions of the thunderclouds seldom extend far into a freezing zone perhaps five miles high.

The higher a temperate-zone thunderstorm extends, the blacker are its underparts, and the more likely it is to unleash severe lightning, torrential rain, destructive winds, and hail. The convergence of two or more storms tends to intensify these factors. Wind-tunnel tests with hail-sized balls have revealed some astounding probable speeds for the updrafts of great storms.

To sustain a one-inch hailstone an updraft of about sixty miles per hour is necessary; a three-inch stone requires a hundred-mile current; and a five-inch stone must have been held aloft at some time by an upblast of two hundred miles an hour! In severe thunderstorms, therefore, 60-mile updrafts (and commensurate adjacent downdrafts) are common, while 100-mile and 200-mile currents are not impossible. Which seems good enough reason for aviators to keep out of them if they can.

An average detached or 'air-mass' thunderstorm starts as a pile of cumulus clouds perhaps five miles or so in diameter, gradually spreading out horizontally and building up vertically as it sucks in warm surface air and gathers to itself near-by clouds. The altitude of its highest cumuliform turrets may be four or five miles in temperate latitudes (or considerably higher in the tropics); and the cirriform anvil or 'cap' may reach six or seven miles aloft.

As the storm moves bodily along with the upper winds, it spreads out laterally to a width of perhaps twenty or thirty miles, so that its whole path resembles a pear in shape. After six or seven hours, with the decline and disappearance of the sun, an air-mass storm usually dies out, having raged in all through a distance of perhaps a hundred and fifty or two hundred miles.

The airplane has of course made it possible to survey the breath-taking topography of thunderstorm clouds, quickly and accurately. One of the TWA 'over-weather' research aviators made a detailed investigation of a large air-mass storm above Missouri. The general storm-top of cumuliform clouds was at twenty-two thousand feet.

The highest cumulo-nimbus turrets boiled up to twenty-five

thousand feet, at which altitude the plane could just clear the turbulent wind-shift line. The cirriform anvil, ten miles wide, extended on up to about thirty-five thousand feet. Flying through the anvil at twenty-six thousand feet, the plane encountered fine snow and some subcooled water droplets which froze as light icing; the temperature was 5° F., compared with 100° F. at the ground.

Even an ordinary thunderstorm has within it a cold front in miniature — a rain-cooled wedge of squall-wind air surging outward and displacing the warm air in front of the storm. But the cold air behind a large-scale front usually comes from arctic or polar regions, whereas the cold air in a thunderstorm is, one might say, manufactured locally by the refrigerating process of the storm. Yet if a very large thunderstorm spreads out laterally beyond fifty miles or so, it begins to take on the characteristics of a line squall.

THE WINGS OF THE STORM

Reduced to typical form and simplest terms, a line squall is nothing more than a violent cold front often spotted with many thunderstorms, continuous or intermittent, along the hundreds of miles of its length. The front itself is necessarily marked by more than ordinarily large temperature contrast between the unlike air masses; by high moisture content and great depth of the warm air mass; and perhaps also by some moisture and instability in the cold air mass.

Also in evidence is the rapid advance of the cold air mass, the enforced retreat (mostly sidewise) of the warm air mass, and conflicting impact between the two. This air-mass conflict shows up as high winds along and across the front, sharply contrasted in direction, and appears on the weather map as isobars packed close together and bending in a sharp V where they cross the front.

From this typical picture as a norm, individual line squalls of course vary widely. Few cold fronts are violent enough to be spectacular line squalls, and few of these, even, conform strictly to type. The one common characteristic of all surface cold fronts, at least over flat country or the ocean, is a marked shift in wind direction, usually from some southerly to some northerly quarter.

The squall line itself usually advances from the northwest or west (less frequently from the southwest) at something like twenty to thirty miles per hour in summer and thirty to forty miles per hour in winter. (Occasionally, in some parts of the country, the squall may come out of the north or even the northeast.)

But rather than further generalities, often violated, here are accounts of two actual line squalls that struck the city of Washington.

One came early in June: At 2 P.M. the sky was still clear, but hazy; and a few clouds were rising in the distant west. The wind was moderate out of the southwest, and some twenty or thirty small sailboats were racing on the wide Potomac River, which was also dotted with various other craft.

Upper clouds gradually grew out of the west, gradually spread and darkened, obscuring the sun in a quarter of an hour. Then a storm collar or squall cloud appeared at lower levels out of the southwest. It consisted of dark gray stratocumulus clouds, whose violent agitation showed the ferocity of their inner winds.

The squall cloud came rapidly on until it passed overhead, but there was no lightning and no rain. The wind slackened somewhat directly under the squall cloud, becoming rather puffy, and the sailboats continued their gay progress all unheeding. But perhaps five minutes after the squall cloud had passed overhead, the surface squall struck very suddenly and fiercely out of the southwest.

As the wind rose shrieking to fifty or sixty miles per hour the sailboats, one after another, were pressed down to the water and capsized, some with sails still set, some under bare poles. The extreme wind lasted only three or four minutes, then moderated to twenty miles an hour for another half-hour, with light rain; after which it fell nearly calm while the rain continued.

Another squall came in May, when the late spring contrasts between tropical and polar air were most intense:

Up to mid-afternoon the sky was perfectly clear but very hazy, the wind moderate southwest. About 3 P.M. the forerunner clouds appeared out of the northwest — perhaps a sort of upper-level squall head — arching across the sky from northeast to southwest. They appeared as a solid wedge of high stratocumulus or low altocumulus, advancing steadily at an altitude between one and two miles.

There were moderate puffs of wind out of the northwest, followed by a gusty northwest wind of twenty-five or thirty miles per hour — but for the moment, no rain. The high clouds swept by overhead and out to eastward, shutting out the sky, and for twenty minutes nothing more happened, except that the pressure jumped up by five hundredths as the cloud edge passed, and the gloom steadily deepened in the northwest.

Then the gray rain curtain appeared, and with it came the real squall at 3.50 P.M. For a few minutes the northwest wind whistled around sixty-five miles an hour, driving torrential rain and rattling hail horizontally against windows and doors. Lightning flickered incessantly above and banged heavily to ground, trees and poles were borne to earth in exposed places, and the horizontal rain flooded over everything, cutting visibility to zero.

The base of the black thunderclouds, when they could be seen at all, looked about one thousand or two thousand feet high. The temperature fell twenty-four degrees in an hour. The worst of the squall was over in a few minutes, but scattered showers and mild thunderstorms continued, interspersed with broken clouds, throughout most of the evening in the unstable invading cold air mass.

These are the picturesque and transient line squalls of spring and summer. In winter, marked cold fronts are less spectacular, but if anything more unpleasant. Their approach is more rapid, and the cold air mass is much colder. Intense winter fronts are likely to be marked by a deep zone of blizzard-like snow.

The worst weight and hazard of summer line squalls comes on the water, as the sailboat episode above indicates.

FLYING THROUGH A LINE SQUALL

Full-bodied line squalls are colorful enough, viewed from the ground or the water. But the way to see and feel one at its best — or its worst — is to get up into the air where we are almost a part of the gyrating squall winds. This is something that wise aviators seldom do, intentionally (except in big, scheduled transports, or in urgent military flying); but they sometimes get caught in squalls or plow through them deliberately in the course of an urgent mission. On the same afternoon that the airship *Hindenburg* crashed in flames at Lakehurst, an army pilot and I were flying from Boston to Washington in *BT*-8 — a medium-fast, stubby, powerful pursuit-type training monoplane:

... Towards Trenton the clouds were more scattered, and we dropped down through them to 2000, heading out westward so as to pass well to windward of Philadelphia's smoke pall. The scattered cumulus clouds, now above us, were becoming larger, with some scattered showers among them. Some of these shining white cumulus heads were beginning to mushroom quite rapidly up into the blue — a sign of thundery weather.

Soon there appeared far ahead, somewhere over northeastern Maryland or southeastern Pennsylvania, a dark line or roll of cloud arching across the entire sky from southwest to northeast. This was the real front, the line squall itself. Great white thunderheads towered along it, in some places much higher than others. Beneath these higher cloud-peaks were gray curtains of rain, and around them flickered the seeming-miniature sparks of lightning. For the time being, the controls were in my not very experienced hands.

Our course was about 240° magnetic (southwest-by-west true), and as we approached to within ten miles or so of the squall line, made out two large thunderheads towering far above the long, continuous roll cloud — turreted white mountains three or four miles high. Under each thunderhead a dark deluge of rain was falling from the squall cloud to the ground. It seemed wise to keep out of these active storms if possible, and we changed course 15° to the right so as to fly midway between them, just under the long, black roll cloud (which appeared fortunately to be well up off the ground, at a height of half a mile or more).

Beneath this rapidly approaching black roof of turbulent cloud the visibility was good — we could see far out ahead into patches of sunlight and shadow beneath other clouds behind the front. We were still cruising in level flight at 2000, and the black ceiling at about 3000, now rushing towards us overhead, seemed to offer plenty of clearance.

As we passed beneath the dark cloud-roof the *BT*-8 began to bounce around crazily. I had expected bumps, but these were so sharp and ferocious that it was all do to maintain on course control of a plane that was, at best, rather tricky to fly. Snatching a glance at the instruments, was alarmed to see the rate-of-climb meter swing to plus 1000 ft/min (12. m.p.h. straight upward), in spite of the fact that was already nosing downward in an effort to get lower and escape from the wildly gyrating black festoons of cloud that seemed now just over our heads.

This black cloud-roof, marked by turbulent downward bulges of mamato-cumulus and momentary 'cloud spouts' or incipient whirlwinds, was getting nearer by the second. It seemed as though we were about to get sucked bodily up into its black chaos. I had the nose shoved down as steeply as seemed prudent, for the air speed was now over 200 m.p.h. and the terrific bumps might strain the wings at this speed. Yet I hesitated to change the throttle setting without instructions from the pilot.

Finally I jammed the nose down in a steeper dive, taking a terrific beating from the bumps and still not losing altitude. At this point the great down-bulging cloud-pocks seemed to be almost couching our heads. Just then the pilot took over, and throttled back to case the bump-strains. And within another minute, we had passed out from under the roll cloud into a region of scattered sunlight and shadow, marked by detached storms in the cold air mass. These detached storms, of course, could be easily flown around.

Passing through a line squall in a controllable airplane may be exciting enough, but it is relatively easy and pleasant compared to navigating through one in an uncontrollable free balloon. Lieutenant Settle of the United States Navy had this unenviable experience over western Pennsylvania on the twenty eighth of May, 1938. Wrote he, in part:

Dropped ballast and at 18.15 (6.15 P.M.) balloon was at 4500 feet altitude; the line cloud overtook and passed under us and we started up rapidly, the variometer going 'hard over' on the ascent side. At 8300 feet I valved two seconds, and balloon leveled off at 8400 feet and then started down at a violent rate, the variometer ink going out of sight on the descent side.

While falling the air was extremely turbulent; hail (size of peas), heavy rain, and snow flurries hit us; the basket swung violently from side to side like a pendulum and spun in azimuth; heavy gusts hit the envelope and basket; the motion of the balloon was like when attempting to hold one on the ground in very gusty surface wind. Lightning and thunder were heavy and apparently right on top of us. The rain and air were very cold and our hands numbed.

During the descent we dropped approximately fifteen bags of ballast; we came through the clouds at 3000 feet and leveled off at about 2000 feet. Found myself under a dense black cloud, the line boundary of which was several miles south of me; the line extended as far as see in either direction.

There were areas of heavy rain along the line on both sides of me and heavy rain to the northward of me, with lightning and thunder on all sides. The rain in my immediate vicinity had temporarily ceased; my air was NNW, about thirty miles per hour, and for a few minutes seemed relatively smooth.

It was soon apparent that we were fast overhauling the line cloud. When nearly up to the line, we again started violently up through the cloud on a second 'vertical circle,' repeating the history of the first one, and attaining a maximum altitude of 8200 feet. Some of the elements were more violent in the second than in the first excursion.

Settle finally managed, by adroit maneuvering with gas valve and ballast release, to case his balloon out of the squall line. He landed at dusk on a farm near Parryopolis. He lived to write of his experience, but others who dared line squalls in free balloons, such as the brilliant American aerologist Meisinger, were less fortunate and plunged to lightning-lit deaths.

Index

F

G

H

I

L

M

P